JC 423 SPA

D1761056

Spaces of Democracy

MAIN LIBRARY
QUEEN MARY, UNIVERSITY OF LONDON
Mile End Road, London E1 4NS
This item is in demand.
THERE ARE HEAVY FINES FOR LATE RETURN

0 NOV 2005

SPACES OF DEMOCRACY

Geographical Perspectives on Citizenship,
Participation and Representation

Edited by
CLIVE BARNETT AND MURRAY LOW

WITHDRAWN
FROM STOCK
QMUL LIBRARY

Ⓢ SAGE Publications
London ● Thousand Oaks ● New Delhi

Editorial arrangement, Chapter 1 © Clive
    Barnett and Murray Low 2004
Chapter 2 © John O'Loughlin 2004
Chapter 3 © Ron Johnston and Charles
    Pattle 2004
Chapter 4 © Richard L. Morrill 2004
Chapter 5 © Sallie A. Marston and
    Katharyne Mitchell 2004

Chapter 6 © David M. Smith 2004
Chapter 7 © Murray Low 2004
Chapter 8 © Lynn A. Staeheli and Don
    Mitchell 2004
Chapter 9 © Gareth A. Jones 2004
Chapter 10 © Clive Barnett 2004
Chapter 11 © Sophie Watson 2004
Chapter 12 © Bryon Miller 2004

First published 2004

Apart from any fair dealing for the purposes of research or
private study, or criticism or review, as permitted under the
Copyright, Designs and Patents Act, 1988, this publication
may be reproduced, stored or transmitted in any form, or by
any means, only with the prior permission in writing of the
publishers, or in the case of reprographic reproduction, in
accordance with the terms of licences issued by the
Copyright Licensing Agency. Inquiries concerning reproduction
outside those terms should be sent to the publishers.

SAGE Publications Ltd
1 Oliver's Yard
55 City Road
London EC1Y 1SP

SAGE Publications Inc.
2455 Teller Road
Thousand Oaks, California 91320

SAGE Publications India Pvt Ltd
B-42, Panchsheel Enclave
Post Box 4109
New Delhi 110 017

British Library Cataloguing in Publication data

A catalogue record for this book is available from the British
Library

ISBN 0 7619 4733 7
ISBN 0 7619 4734 5 (pbk)

Library of Congress Control Number 2004102659

QM LIBRARY
(MILE END)

Typeset by C&M Digitals (P) Ltd., Chennai, India
Printed in Great Britain by Athenaeum Press, Gateshead

Contents

List of Contributors

Clive Barnett is Lecturer in Human Geography, Faculty of Social Sciences, Open University, Milton Keynes.

Ron Johnston is Professor of Geography, School of Geographical Sciences, University of Bristol.

Gareth A. Jones is Senior Lecturer in Development Geography, Department of Geography and Environment, London School of Economics and Political Science.

Murray Low is Lecturer in Human Geography, Department of Geography and Environment, London School of Economics and Political Science.

Sallie A. Marston is Professor of Geography, Department of Geography and Regional Development, University of Arizona, Tuscon.

Byron Miller is Associate Professor and Director of the Urban Studies Programme, University of Calgary, Alberta.

Richard L. Morrill is Professor Emeritus, Department of Geography, University of Washington, Seattle.

Don Mitchell is Professor of Geography, Department of Geography, Syracuse University, Syracuse.

Katharyne Mitchell is Associate Professor of Geography, Department of Geography, University of Washington, Seattle.

John O'Loughlin is Professor of Geography, Department of Geography, University of Colorado, Boulder.

Charles Pattie is Professor of Geography, Department of Geography, University of Sheffield.

David M. Smith is Professor Emeritus, Department of Geography, Queen Mary, University of London.

Lynn A. Staeheli is Professor, Department of Geography, University of Colorado, Boulder.

Sophie Watson is Professor of Sociology, Faculty of Social Sciences, Open University, Milton Keynes.

Acknowledgements

This volume grew out of sessions at the Annual Meetings of the Royal Geographical Society–Institute of British Geographers at The University of Sussex in Brighton, and the Association of American Geographers in Pittsburgh, both in 2000. We would like to thank all those who took part on those occasions. We would also like to thank Drew Ellis and Jonathan Tooby of the School of Geographical Sciences, University of Bristol, for preparing the maps and figures. We would like to thank Robert Rojek at Sage for his help during the preparation of the book. Finally, we would like to thank Julie McLaren and Abbey Halcli for their support and encouragement throughout.

Clive Barnett and Murray Low

1 Geography and Democracy: An Introduction

Clive Barnett and Murray Low

Where is Democracy?

Amid debates about globalization, neo-liberalism, and anti-capitalism, it is easy to forget that probably the most significant global trend of the last two decades has been the proliferation of political regimes that claim to be democracies. Democracy refers to the idea that political rule should, in some sense, be in the hands of ordinary people. It is also a set of processes and procedures for translating this idea into practices of institutionalized popular rule. In a remarkably short space of time, commitment to democracy has become near universal. The universalization of democracy as an ideal, if not as a set of agreed-upon practices, is historically unprecedented: 'Nothing else in the world which had, as far as we can tell, quite such local, casual, and concrete origins enjoys the same untrammeled authority for ordinary human beings today, and does so virtually across the globe' (Dunn, 1992: 239). This assertion pinpoints one key geographical dimension of the contemporary ascendancy of democratic norms. This is the problematic relationship between the particular historical-geography of democracy's 'origins' on the one hand, and democracy's more recent globalization on the other. However, it is striking how little impact processes of democratization, or democracy as a broader theme, have had on research agendas in human geography. While a great deal of critical analysis is implicitly motivated by democratic norms, there is relatively little empirical research or theoretical work that explicitly takes democracy to be central to the human geographic endeavour. This book aims to address this lacuna, by bringing together contributions from across the discipline of geography, addressing various research fields in which democracy is often a veiled backdrop, but not usually a topic of explicit reflection. We hope the book will thereby help to encourage the sort of detailed attention to issues of normative political theory that has recently been called for by others (Agnew, 2002: 164–78).

The ghostly presence of democracy in geography can be illustrated with reference to a number of fields. First, debates on the geography of the state, starting in the 1970s with Marxist-inspired work on the capitalist state, and developing in the 1980s and 1990s through an engagement with regulation theory, certainly took the concept of legitimacy and the representative dimensions of state institutions into account. However, detailed examination of routine democratic procedures of participation and representation have remained peripheral to the analyses developed in this area, which remain constrained by a conceptualization of political processes as derivative of more fundamental economic interests. More broadly, the neo-Gramscian state theory most favoured in geography has remained largely untouched by the flowering in the last three decades of post-Enlightenment liberal political philosophy that has reinvigorated debates about democracy, citizenship, and power.

The concern with social justice stands as a second example of the marginalization of democracy as a theme in human geography. This might sound counter-intuitive, since the value of democracy as a form of rule is often linked to its role in securing social justice (Rawls, 1971). Geographers have engaged in debates about social justice since the 1970s. But geographers' interest in these questions has tended to focus on substantive distributive outcomes and spatial patterns, rather than on the issues of political process and procedure that would lead to democracy becoming a central topic for debate. Themes of geography and justice have been revitalized recently by the development of an explicit concern with moral and ethical issues (see Proctor and Smith, 1999). Yet the focus of this ethical turn has been on moral rather than political theory, leading to a concentration on questions of ethical responsibility detached from both wider issues of institutional design and political processes.

A third example of the displacement of democracy in geography is recent research on the geographies of citizenship. This work has concentrated on relationships between migration, citizenship and discourses of belonging and identity, and how these shape differential access to material and symbolic resources from states. Most discussions of these matters in geography have been conducted in light of the question of whether globalization complicates the spatial dimensions of membership and access to material resources of citizenship. The uneven development of rights of *political* citizenship, and the practices of mobilization and engagement these enable, has received relatively little direct treatment by comparison (Low, 2000). Electoral geography is the area of human geography research that has consistently addressed the political and participatory dimensions of citizenship rights, and by extension the area that has been most consistently focused on core features of democratic politics. An interest in the dynamics of democratic process and procedure has been unavoidable in this work, as has a focus on questions about political representation. While there are many empirically detailed analyses of electoral 'bias' in particular political systems, the broader issues raised by the subject matter of electoral geography have often remained unexplored. Only recently have geographers begun to explore the links between this predominantly quantitative-empirical field of research, and normative issues of political theory and democratic justice (Johnston, 1999; Hannah, 2001).

Finally, one might expect that the proliferation of culturally-inflected research in human geography would have been the occasion for a more systematic engagement with political theory. Power has become a ubiquitous reference point in the new cultural geography, and in work touched by the cultural turn more widely (Sharp et al., 1999). However, on closer examination, this concept is a conceptual black box rarely opened up to detailed analysis (see Allen, 2003). Too often, the recourse to the vocabulary of resistance and hegemony in cultural theory marks the point at which reflection on first principles is displaced in favour of the imaginary alignment of the academic analyst with popular struggles (see Barnett, 2004).

These examples point towards a recurrent preference in human geography for the language of explanatory rigour, social change, or policy relevance, rather than reflection on normative issues. As a consequence, geography's treatment of politics is characterized by *theoreticism*. By this we mean a tendency to deduce desirable political outcomes from deeper interests, established outside political processes, into which the academic researcher has a privileged insight. This preoccupation is often combined with voluntaristic injunctions to the community of researchers, governments, or social movements to work to help bring these outcomes about. In short, the very terms in which geographers have engaged in discussion of politics, justice, citizens, elections, have nourished an avoidance of reflection on the normative presuppositions of political institutions and on the basic criteria of political judgement underpinning democratic processes – criteria concerning what is right, what is just, what is good, and concerning *how* best to bring good, just, rightful outcomes about.

As other commentators have argued (Sayer and Storper, 1997; Corbridge, 1998), radical traditions of geographical research have persistently evaded normative political philosophy in favour of either the abstracted-individualism of ethical reflection or the certainties of radical political critique. It is in areas of the discipline often thought of as more 'applied' that one can find the most sustained reflection on the normative issues raised by democratization processes. This is the case, for example, in both urban planning and environmental policy studies, in which the meanings and practicalities of deliberative decision-making and participatory democracy have been extensively discussed (e.g. Burgess et al., 1998; Hajer and Kesselring, 1999; Mason, 2001; O'Neill, 2001; Owens, 2001). Likewise, it is among development geographers that one finds sustained critical discussions of the concepts of civil society and social capital, and of the meanings of participation, representation and empowerment, all issues with implications and currency far beyond the Global South (e.g. McIlwaine, 1998; Jeffrey, 2000; Mercer, 2002; Williams et al., 2003). Planning studies, environmental studies and development geography all connect up with broader interdisciplinary arenas where issues of democratic theory have been central in shaping research agendas. This is less true of the favoured interlocutors of 'mainstream' critical human geography.

The disconnection of a theoretically confident tradition of critical human geography from the concerns of political philosophy and democratic theory requires some explanation. Is it because these other fields are not sophisticated enough in their treatment of space, spatiality, or scale to satisfy the agenda of critical human geography? As we will argue below, this explanation does not stand up to scrutiny. In order to explore the question further, we want to identify three points of potential overlap but actual separation between geographical research and democratic theory. First, there is the problematic status of liberalism in human geography. We relate this issue to geography's treatment of the state. Secondly, there is the question of the degree to which the geographical imaginations of human geography and political theory diverge. Thirdly, there is the thorny problem of how to understand the value of universalism, a concept that is central to debates about democracy, but which geographers find hard to assimilate to their disciplinary matrix of ideas. In flagging these three themes, we want to contextualize the chapters in the book, by providing some sense of the most fruitful cross-disciplinary engagements towards which they might lead.

Rehabilitating Liberalism

The templates for democratic institutions in the West, and indeed in most other contexts today, are usually referred to as being liberal in character. Alternative conceptions of democracy (including communitarian, deliberative, participatory, radical, and discursive approaches) all tend to define their own virtues by reference to the strengths and weaknesses of liberal theory and practice. However, liberalism is a rather broad label for a heterogeneous collection of ideas and practices. One tradition of liberalism, best exemplified by Hayek, explicitly seeks to restrict the scope of democratic decision-making in the name of the higher goods of personal liberty and free markets. One irony of the ubiquitous recourse to the vocabulary of 'neo-liberalism' in contemporary left-critical discourse is, however, the identification of liberalism *tout court* with this particular variety of conservative political thought. In this unlikely convergence, liberalism is reduced to a doctrine that counterposes the state to the market.

This mirroring of left and right readings of classical liberal doctrine erases the historical variety of *liberalisms* (Gaus, 2003). The market liberalism exemplified by Hayek echoes a broader discourse of elitist disenchantment with mass democracy, which includes Weber, Pareto, Schmitt, Michels, and Schumpeter. What connects these thinkers is an intuition that the mass scale of modern polities, in both spatial and numerical terms, renders democracy implausible and hazardous. By contrast, there is a diverse tradition of avowedly liberal thought that reasserts the plausibility and value of extending democratic procedures across larger scales and into a wider range of activities. This tradition would include the work of Robert Dahl, John Dewey, Otto Kirchheimer, Carole Pateman, and John Rawls, as well as that of Noberto Bobbio, Jürgen Habermas, Hannah Pitkin, and Roberto Unger. This is a disparate group, but that is partly our point. It comprises a range of different projects that include a revivified Kantian republicanism, political liberalism, civic republicanism, and democratic liberalism. The key feature that these projects share is an effort to overcome ossified dualisms between equity and liberty, by finding practically informed ways of thinking through disputed conceptions of the right, the good, and justice. Taken together, these post-Enlightenment liberalisms can be said to constitute a broad tradition of radical democracy, one that is characterized above all by a shared concern with defining democracy in relation to practices of citizen participation.

We think it important to reassert the significance of this tradition of self-consciously egalitarian, democratic liberalism precisely because liberalism largely remains a denigrated tradition of thought in critical human geography. Radical human geography explicitly emerged by turning its back on liberal approaches in the 1970s. One consequence of this has already been noted. This is the persistent tendency to elevate explanatory accounts of socio-spatial process and substantive (outcome-oriented) accounts of justice over an engagement with the significance of procedural issues of participation, representation, and accountability (see Katznelson, 1996). As Howell (1993: 305) has observed, while geographers have engaged with an ever-widening range of theoretical ideas, the dimension of normative reflection on political principles contained in writers such as Habermas, Foucault, or Derrida is too often obscured 'by the use to which they are put [...] as part of a generic *social* theory to which we as geographers appeal almost exclusively for validation'. This preference for social rather than political theory means that it is rare to find discussions of the geographical dimensions of inequality, or the spatialities of identity and difference, which are able to address fundamental questions concerning the significance of the values of equality, diversity, or difference that such analyses implicitly invoke.

The suspicion of liberal traditions of political theory has had two further consequences for the ways in which geographers address themes of democracy. First, liberalism as political theory is easily associated with the manifest flaws of 'actually existing democracies'. It is certainly true that elements of liberal discourse (rights, freedom, liberty) can readily take on ideological value in defending undemocratic or illiberal practices. But this is not a unique feature of liberalism. In fact, this ideological potential seems a very good reason for critically reconfiguring key terms such as 'rights', 'liberty' or 'representation', rather than assuming that they cannot be divorced from compromised realities and that we must find less tainted images of authentic political action.

This brings us to our second point, which is that ideal-typical liberal theories of democracy are persistently framed as the benchmark against which truly radical theories of democracy should be judged. As a result, the definition of radical politics is moved further and further away from the sites of mundane politics. Of course, one of the crucial insights provoked by a variety of new social movement mobilizations since the 1960s is the political stake involved in distinguishing what is politics from what is not. It is often argued that this requires that the meaning of 'the political' should be reframed beyond narrowly defined understandings of government, constitutional rule, voting, or party support. One example is Laclau and Mouffe's (1985) conceptualization of radical democracy. This is perhaps the most important example of political theory to attract sustained attention in human geography (see Jones and Moss, 1995; Brown, 1997; Robinson, 1998). The characteristic Marxist response to their distinctive poststructuralist, post-Marxism has been to dismiss it as revised liberal pluralism. However, in their concern to destabilize standard conceptions of interests, the people, or representation (and to develop an alternative vocabulary of articulation and antagonism), it is clear that Laclau and Mouffe are strongly committed to moving decisively beyond liberal formulations of democracy.

Counterposing mere 'politics', with all its disappointments and limitations, to the question of 'the political' is central to the poststructuralist project of radicalizing democracy. It is associated with the claim that grasping the essence of the political requires a form of analysis utterly different from liberal rationalism, which is supposedly unable to acknowledge irreducible conflict and antagonism. But this leads poststructuralist accounts of radical democracy into the rather thankless task of trying to redeem some democratic value from the plainly anti-democratic political thought of writers such as Martin Heidegger or Carl Schmitt. With their analytics of forgetting and disclosure, neutralization and depolitization, these writers have become the unlikely foundation for new formulations of radical political action that apparently escape the inauthenticities of ordinary politics. In this strand of work, the banality of ordinary politics is transcended by the promise of a more heroic variety of political transformation rooted in an image of liberating the creativity unjustly contained by the limits of state, capital, or bureaucracy. So it is that poststructuralist accounts of the political come to resemble a form of idealistic superliberalism (Benhabib, 1992: 16). They claim to be more pluralistic, tolerant, and affirmative of difference than conventional liberalism, yet are unwilling to acknowledge the practical dependence of these values on the real achievements of liberal political cultures. This in turn explains the consistent difficulty that poststructuralist theories have in accounting for democracy as a specific sort of institutionalized politics (Dietz, 1998; see also Amin and Thrift, 2002), beyond modelling political action on aesthetic practices such as performing or reading.

The poststructuralist reconstruction of radical democracy therefore illustrates the paradox of the idea of cultural politics more generally. This idea carries a double resonance, broadening the range of activities understood to be in some sense political, but at the same time it carries the risk of jettisoning any concern for the realms in which politics most obviously still goes on. The danger lies in presuming that a whole set of traditional problems in democratic theory – the nature of representation, the meaning of legitimacy, and so on – can be easily resolved. With the near universalization of democracy in both theory and practice, the attention of critical analysis has shifted away from justifying democracy against other forms of political arrangement, towards finding fissures at the margins of actually existing regimes that promise better forms of democracy. As we suggest below, this dynamic of perfectibility might well be a distinctive feature of democracy as a regime of rule. But one unforeseen consequence of this democratically-oriented critique of actually existing democracy is a tendency to always assume that 'democratic discontent emerges from the institutions of representative democracy and can best be ameliorated by the wider democratization of social relations as they are reproduced in civil society' (Squires, 2002: 133). This stark opposition between representative forms of democratic politics, presumed to be the source of dissatisfaction, and idealized models of alternative politics, leads to an underemphasis on the changing dynamics of formal political institutions of the state. This tendency is exemplified by recurrent calls in political geography to transcend 'state-centric' views of politics (see Low, 2003). Suspicion of the state as a central object of geographical concern is rationalized in terms of facing up to the historical and geographical specificity of state forms, and by calls for thinking about the possibilities of organizing politics differently (Taylor, 1994; Agnew, 1998).

The suspicion in geography of state-centred understandings of politics is the main reason for the persistent non-engagement with liberal political theory. Liberalism is marked by a double recognition of the unavoidability of centralized decision-making and a resolute suspicion of its hazards. This implies that democracy needs to be understood in *relational* terms, as a means through which autonomous actors engage with, act for, influence, and remain accountable to other actors, a process carried on through institutional arrangements that embed particular norms of conduct. Two-dimensional political imaginations of resistance or hegemony are rather limited in their understanding of contemporary forms of protest, campaigning, and dissent, in so far as they tend to underplay the commitment to engaging with centralized forms of power, both public and private, that most often distinguishes contemporary social movements. Rather than resistance and hegemony, perhaps the better master-concept for understanding such politics is that classically liberal motif of opposition. Even the most radical forms of contemporary political action are animated by democratic demands (that decisions should be made out in the open and should be based on consent, and that institutions and organizations should be accountable), underwritten by democratic principles (above all, that the legitimacy of rule depends on authorization by ordinary people affected by the consequences of actions), and employ strategies that are the stock in trade of democratic social movement mobilization stretching back two hundred years (the theatrical mobilization of large numbers of supporters in public spaces). At the same time as appealing to the idea of democracy's perfectibility, these mobilizations for greater democracy testify to the impossibility of any established set of democratic procedures ever completely embodying the preferences of all in an unambiguously fair manner (Shapiro, 1999: 31). If, then, democratic politics requires opportunities for inclusive participation and accountable representation, then the value of these is only realized in the context of robust and varied practices of opposition (ibid., 31–45, 235–8).

In short, the diverse heritage of liberalism is too important to be dismissed by those interested in progressive social change. It remains an essential reference point for connecting the actualities of political action to reflection on the principles and procedures that define democratic justice. It is this space that is closed down by market-based models of democratic choice, as well as by agonistic models of contingent identifications expressed in insurgent acts of resistance. It is, moreover, important to redeem the term 'radical democracy' from a narrow understanding of identity-politics. Rehabilitating the emphasis found in the use of this phrase by both Dewey and Habermas, radical democracy refers to an expansive sense of politics, involving participation in a range of formal and informal practices of identification and opinion-formation combined with a pragmatic orientation towards getting things done. While keeping open questions about the status and scope of political action, it also suggests a less distanced engagement with what is ordinarily defined as 'politics' – with matters of policy, legislation, parties, lobbying, organizing – than is often countenanced in more rarefied accounts of radical counter-hegemonic politics.

The key theme linking the alternative liberalisms we sketched at the start of this section is a focus on the 'how' of power. Rather than presuming that political judgement is reducible to a question of who holds power or of which forces are in the political ascendancy, an emphasis on procedural forms of power focuses upon the difference that exercising power in relation to practices of publicity, justification, and accountability makes to the substance and quality of outcomes (Habermas, 1996). The emphasis of the broad tradition of participatory radical democracy upon the combination of citizen participation and decisive action opens up issues of political judgement to resolutely geographical forms of interpretation. This is not least the case in so far as the relationship between democratic participation and democratic decision turns on a paradox of scale – on the problem of how to institutionalize effective citizen participation in functionally complex, socially differentiated, and spatially and numerically extensive societies. Ideas of participatory radical democracy, understood as a distinctive variety of post-Enlightenment liberal political theory, therefore require a reconsideration of the distinctive imaginary geographies of modern democratic theory.

Imaginary Geographies of Democratic Theory

Democratic theory has had a persistent problem addressing the significance of its own implicit geographical assumptions. This is particularly the case with respect to the conceptualization of borders and boundaries (see Taylor, 1994, 1995; Anderson, 2002), a key issue in determining the identity and scope of democratic political rule. Very often, geographical assumptions of bounded territorial entities are not thematized in democratic theory, although there is also a stronger positive argument to the effect that democracy is not possible without sharp geographical boundaries between polities. While acknowledging the problematic elements of political-theoretical assumptions about the geography of democracy, we also want to suggest that the predominant geographical imagination shaping research agendas in human geography might lead to potential points of connection with democratic theory being by-passed. Geographers' entry point into wider interdisciplinary debates has been their specialization on space, place, and scale as objects of analysis. However, this might also serve as a barrier to certain forms of interaction. There are three dimensions to this claim. First, geographers' conceptualizations of space, place and scale emphasize complexity and differentiation. Geographers' spaces are uneven, relational, reticulated, blurry, stratified, striated, folded over, porous, and so on. Secondly, geographers' strong emphasis upon the constructed, non-natural qualities of territorial entities has led to a wariness of focusing on national scales of political action. There is an in-built impetus to de-centre and de-naturalize the national scale as the privileged focus of attention. This can lead to a further displacement of much of the most routine and ordinary activity of everyday democratic politics already encouraged by poststructuralist understandings of radical democracy. Thirdly, and following from both of these previous points, the preferred scales of analysis for geographical research tend to be both above and below the nation-state, with the local, the urban, the regional, and the transnational. Even if territorial notions of multiple scales are rejected as overly formal and constrictive, then the effect is still to emphasize a further complication of flows, connections, networks and fluidities, (Amin, 2002). Combining these three observations, one might conjecture that a justified conceptual hollowing-out of the nation-state as the taken-for-granted scale of political analysis easily leads to an automatic presumption against national-level forms of political practice. This supports an unexamined prejudice against some of the most mundane elements of liberal representative democracy, which are again reduced to the benchmarks against which more radical understandings of democracy will be constructed.

The tension between the conceptual emphasis upon re-imagining spatial complexity on the one hand and the embedded geographies of democratic politics on the other is not only a problem for geography. It generates recurrent problems for political theorists of democracy themselves. The disconnection between geography and political theory can no longer be ascribed to the claim that democratic theory is inadequately sensitive to the spatialities of social processes. Modern political theory has, in fact, always been concerned with the difference that geography makes to the qualities of democratic rule. This is the case with theorists as diverse as Montesquieu, Rousseau, Madison, Burke, Paine, Tocqueville, Condorcet and Constant (Manin, 1997), through to twentieth-century political science preoccupations with democracy and size (see Dahl and Tufte, 1973; Dahl, 1989). Furthermore, there has been a veritable 'geographical turn' in recent political philosophy and international relations theory. This would include the deconstruction of the imaginary geographies of international relations theory (Connolly, 1991; Walker, 1993) that connect with geographers' own critiques of the so-called 'territorial trap' (Agnew, 1994; Low, 1997). The supposedly taken-for-granted nature of boundaries and national-level processes has clearly had its day in political theory (Shapiro and Hacker-Cordón, 1999). Geography has also 'broken out' in debates about the scope of political communities and political obligations sparked by ongoing confrontations of liberal and communitarian political imaginaries (see O'Neill, 2000). One central context for these debates is the process of transnational migration, which has provided a real world reference point for questioning the taken-for-granted spatial assumptions grounding modern understandings of popular democratic legitimacy. The idea that citizens are obliged to respect the legitimacy of laws by virtue of having participated in making them has been questioned on the grounds that it unreasonably stakes political community upon shared cultural identities located within clearly delineated territories (Cole, 2000).

These developments in turn inform discussions about the value of national identity as the necessary prerequisite of citizenship (see Honig, 2001; Benhabib, 2002), which develop the revival of interest in the Kantian theme of cosmopolitanism and hospitality in the work of both Derrida (2001) and Habermas (1998). These debates have coincided with critical geographical work that more explicitly addresses the assumptions about geography, space, and place built into abstract formulas of cosmopolitan ethics and politics (Entrikin, 1999; Harvey, 2000). Other areas in which the geographies of democratic theory have been conceptually twisted and stretched include consideration of the difference that geographical scale makes to the possibilities of instantiating democracy at the level of the European Union (Schmitter, 1999), and in ongoing work on the role of social movements in historically consolidating national territorial democracies (Hanagan and Tilly, 1999). In this latter area there is an explicit and critical reflection on the centrality of questions of space to the ways in which social movements are organized and develop (Sewell, 2001), an interest that connects up with the growing interest in human geography in the spatialities and scales of social movement activism.

This increasing focus among political theorists on issues of space, scale, borders, and boundaries suggests that there is considerable scope for a productive engagement with geography over issues of shared concern. But it also indicates that this engagement cannot plausibly take the form of geographers supposing that they have a monopoly on the most innovative ways of thinking about space, scale and territory. Dialogue would be better facilitated by a shift in the balance and rationale of geographers' arguments, with rather less focus on complicating understandings of space, and more on theorizing and investigating the reconfiguration of inherited geographies of democracy within a converging intellectual field where asserting that 'geography matters' is no longer an issue.

However, there might be a more fundamental tension at stake between the two disciplinary fields of human geography and political theory than their different conceptualizations of space and territory. Political theory's traditional investment in taken-for-granted geographical dimensions of democratic political action, or its preoccupation with relatively simple concepts of scale and geographically contained polities, is not simply a conceptual blind-spot. It might stem from a fundamental investment in the value of universalism in defining the value of democracy. Squaring this commitment with the actualities of worldly difference tends to be achieved by holding fast to notions of bounded political entities within which universal rights and obligations are ideally secured. In the wake of theoretical and political criticisms that affirm difference and diversity over false universalism, this investment might be at odds with geography's already deeply ingrained preference for the value of the particular and the specific.

Spaces of Difference and Universalism

We have argued that the conceptual and polemical trajectory of critical human geography has led to a search for politics away from the most obvious site of democratic contention (i.e. the state), and has favoured ways of understanding political processes which reject the starting points of the tradition of thought in which the meanings of modern democracy have been most systematically subjected to normative-conceptual analysis (i.e. liberalism). In turn, we have suggested that geography's disciplinary concern with the complexity of spatial and scalar relations sits uneasily with the characteristic ways in which space, scale, and territory have been conceptualized in democratic political theory, although there may be signs of a convergence of interest in this respect. It is the combination of these two emphases – the suspicion of state-centred, liberal political theories, and the attraction to ever more complex understandings of space and scale – that explains the strong affinity that geography has expressed with theoretical critiques of universalizing democratic theory made in the name of difference, diversity, and otherness. It is a commonplace to observe that liberal political theories have difficulty accommodating difference and pluralism at a theoretical level (see Young, 1990; Phillips, 1991; Mouffe, 1998). And it is a short step from this philosophical critique of concepts of identity and difference to the claim that liberalism fails to address geographical variations in socio-cultural and political arrangements. However, these two arguments – about worldly differences between peoples, places, and polities on the one hand and about the conceptualization of difference as a philosophical, ethical, and political value on the other – might not be so easily, or wisely, aligned as is sometimes supposed.

The fundamental question facing any critical analysis of democracy is whether or not the claims of universality built into democratic theory are nothing more than culturally specific norms. This is not simply a question of whether particular procedural models of democracy are appropriate as global norms. It is also a question of whether the models of universal interest and binding obligation that underwrite modern democratic theory might in fact operate to reproduce systematic, hierarchical exclusions and inequalities. A fundamental critical task is to unravel the logical and normative relations between the *genesis* and *form* of modern democracy. Does the historical geo-graphy of actually existing democracy mean that democracy, as a value, is inherently 'Western' in its essence? Some writers argue that the so-called 'third wave of democratization' in the last three decades is indeed the realization of a historical teleology towards liberal representative democracy (e.g. Fukyama, 1993). In this sort of narrative, democracy is assumed to be a cultural formation with characteristics that are distinctively 'Western' (e.g. Spinosa et al., 1999). These sorts of assumptions are in turn countered by the charge that the universalism of liberal democracy is a false one, covering over particularistic exclusions (Parekh, 1993), and that the spread of democratic governance is as much a reflection of the post-Cold War geopolitics of donor funding, good governance, and brokered democratic transitions.

Neither position is really adequate, since neither addresses in detail the disjunctive relationship between what might be called democracy's 'context of discovery' and its 'contexts of justification'. Discussions of the meaning of democracy, whether by champions or critics, too often simply assume the identity of democracy as Western, and in turn conflate the significance of universalistic normative procedures with particular cultural norms of conduct and aspiration (see Sen, 1999a). But democratization, both historically and in the present period, has had multiple trajectories. In this respect, Schaffer's (1998) analysis of the practice of democracy in modern Senegal is notable for its recourse to the thematic of translation in understanding the cross-cultural variability of democratic norms. Schaffer underscores two points: first, that the meanings ascribed to democracy vary across cultures and contexts, but without losing their universal resonance; and secondly, democracy emerges as a modality of rule that emphasizes talking, agreeing, arguing, dissenting, getting things done, and holding to account. This analysis underscores the sense that democracy is the name for variable forms of rule that fold together diverse interests and plural identities in a pattern of decisive action in which the norm of ordinary people participating in the actions affecting them is accorded priority.

The argument that democracy's meaning is historically and geographically variable, without being wholly indeterminate, is the theme of David Slater's (2002) recent critique of Eurocentric discourses of democratization. Slater is keenly aware of the unequal geopolitics of the diffusion of democracy, but is equally keen to stress that this does not de-legitimize democracy as a goal or form of politics. By excavating alternative, non-Western traditions of democratic theory and practice, this sort of self-consciously post-colonial critique of theories of democratization demonstrates that actual processes of political transition are likely to be the outcome of contingent combinations of 'top-down' international pressures for good governance and 'bottom-up' pressures for social change and greater accountability.

Following Slater, we want to suggest that any either/or choice regarding democracy needs to be resisted. Treating liberal democracy as either irredeemably parochial or as undifferentiated in its universal application is premised on an image of cultural space in terms of bounded containers, a spatial imagination from which the opposition between universalism and relativism is in large part derived (see Connolly, 2000). As a way out of the oppositional polemics that surround discussions of democracy's origins and application, we think it might be useful to consider the different trajectories of democratization in terms of *family resemblances*. This idea follows from the observation that democratization often involves a combination of distinctively local features, appropriations from elsewhere, and new inventions. For example, the emergence of modern democracy in the eighteenth-century depended on the appropriation of pre-democratic political mechanisms like representation (Manin, 1997). In turn, twentieth-century anti-colonial movements borrowed and re-invented nationalist discourses, in the process establishing the value of national, sovereign independence as a basic element of modern understandings of democracy (Held, 1997). And this hybridization of democracy is increasingly institutionalized through organizational networks of policy advocacy, social movement mobilization, and human rights monitoring.

These ideas – that democracy is a necessarily plural form, one that moves through processes of translation, and that different variants are related according to different degrees of family resemblance – allows us to specify the geographical significance of thinking of democracy as an 'essentially contested concept' (Connolly, 1993: 9–44). To describe democracy in these terms is not merely to suggest that people disagree about the meaning of the term. More fundamentally, it suggests that this disagreement is structured around recurrent contradictions between essential elements of the term – for example, between individual liberty and collective action, between majoritarian principles and minority rights, between participation and delegation. Democracy is essentially contested because it is an inherently appraisive category – people are concerned with deciding the degree to which particular situations are more or less democratic. And crucially, democracy is also essentially contested because the positive appraisal of a context as democratic includes within it an allowance for changing circumstances and modifications (see Gallie, 1956: 183–7). This means that the precise form of democratic rule cannot be established in advance, but is open to modification in light of new circumstances. Thinking of the universality of democracy in terms of family resemblance, hybrid appropriations, and inventive translations underscores the extent to which the problem of applying practices and norms developed in one context to new contexts is at the root of the critique of democracy's presumptive universalism. And this implies that the conceptualization of democracy, and not just its empirical investigation, is an inherently geographical enterprise.

Whatever their origins, discourses of democracy, citizenship, and human rights now form an almost ubiquitous formative-context for political action by states, corporations, popular movements, or individual citizens. This observation is not meant to endorse a complacent understanding of democracy as benignly capacious, but rather to emphasize the extent to which the normative horizon of the discourse of democracy shapes real-world conflicts. This allows us to understand the positive attraction (as distinct simply from a negative critical force) of the difference-critique of universalism. This critique is most often articulated in a register that appeals, at least implicitly, to norms of universality and equity that it finds to be contravened in practice. The critique of democratic universalism made in the name of the cultural relativity of values re-inscribes rather than rejects universalism: 'The meaning of the relative does not erase, but rather carries within it, a universal exigency' (Lefort, 2000: 144). Critiques of false universalism are made in the name of the equal recognition of identities, or of equal respect for competing notions of the good life. This observation does not negate the force of the critique, in the manner of a liar's paradox. Rather, it suggests a different alignment of the universal and the relative, not as polar opposites, but as different registers of judgement.

Our argument is, then, that the difference-critique of liberalism does not have direct political relevance as such, but rather functions as a supplementary critique that calls for certain principles and practices to be reconfigured in new ways. Chief among these is universalism, the value of which needs to be recast. There are two broad approaches to the post-Enlightenment revision of universalism in the wake of the difference-critique of liberalism. These two approaches – one of which involves a commitment to minimal universalism, the other a rethinking of universalism as an orientation towards openness to otherness – share in what Stephen White (2000) refers to as a commitment to 'weak ontology'. That is, they are approaches that affirm certain fundamental values while at the same time acknowledging the contingency and contestability of those fundamentals.

The first of these approaches to rethinking the value of universalism follows from the observed similarities in the meanings ascribed to democracy in variable historical, geographical, and cultural contexts. This is used as a basis for affirming a base-level, minimal universalism in defining human needs, capabilities, and standards of justice (see Corbridge, 1993: 1998). This is an argument most coherently developed in the work of Amartya Sen (1999b) and Martha Nussbaum (2000), both of whom argue for a universalism of basic human capabilities. Their position gives considerable importance to the idea that a key human good is the practice of asking questions and offering justifications through which human needs are defined. Drawing on a similarly Kantian heritage, Onora O'Neill (2000) deduces a universalism premised on practical actions which are stretched out over space and time, and which implicitly assign competency, agency, and equal moral respect to others irrespective of their ascribed identities.

The second approach to recasting universalism is distinct from the post-foundational philosophical anthropologies implied by the adherence to a minimal universalism of reasonably defined needs. In this second approach, the critique of static, essentialist universalisms of justice, democracy, or rationality leads to a rein-terpretation of universalism in terms of openness to otherness. The deconstruction of exclusionary universalism leads to a redefinition of universality not as a singular, converged set of values (being-the-same), but in terms of being-together (see Nancy, 1991). From this perspective, the value of universalizing discourse lies less in its descriptive content than it does in the implied commitment to listen to and respond to claims for justice from others that is implied by invoking a universalist register. This argument is developed, for example, in Iris Young's (1993) concep-tualization of communicative democracy, in which democratic justice does not presume the transcendence of particularity in favour of a shared universal perspec-tive. It depends instead on a shift from a self-centred understanding of needs to the recognition of other perspectives and a commitment to negotiation. 'Appeals to justice and claims of injustice [...] do not reflect an agreement [on universal principles]; they are rather the starting point of a certain kind of *debate*. To invoke the language of justice and injustice is to make a *claim*, a claim that we together have obligations of certain sorts to one another' (Young, 1998: 40). In this formula, universality is rethought not in terms of sameness, but in terms of openness. Openness is a value that presupposes plurality not sameness. This recasts rather than rejects the value of the universal, understood as an aspiration or impulse towards which claims for justice are oriented without presuming that this requires complete transcendence of partial positions.

This second approach to the universalism of democracy points towards the distinctive temporality that is characteristic of democratic rule. If democracy is understood to have had no essence (which is not the same as saying it is a purely empty category), this is because democratic rule is oriented towards the future. It is a form of rule that anticipates revision. In an abstract register, this is the sense of Derrida's (1992) account of 'democracy to come', which turns upon two notions of the future: the future as programmed and planned; and radical openness of the future as the wholly unexpected, what cannot be anticipated. Derrida suggests that the promise of democracy inheres in the relationship between these two temporal registers: 'For democracy remains to come; this is its essence in so far as it remains: not only will it remain indefinitely perfectible, hence always insufficient and future, but, belonging to the time of the promise, it will always remain, in each of its future times, to come: even where there is demo-cracy, it never exists, it is never present, it remains the theme of a non-presentable concept' (Derrida, 1997: 306). This philosophical understanding of the temporality of democracy's promise of perfectibility connects to a more pragmatic observation concerning the basic mechanisms of democratic modes of rule. Regular elections, rights to free assembly, and so on, all embody a commitment to deal with irreconcilable difference and unstable identifications in a peaceable fashion by *temporizing* conflicts. This depends on institutionalizing a distinctive temporal rhythm that combines open-ended deliberation, temporary identifications, the punctuality of decisive action, and retrospective accountability (Dunn, 1999). Democracy, in short, is a political form that enables action that is being decisive without being certain, and is therefore open to contestation and revision. And this implies that it is important not to think of democracy in terms of *identity*, whether this refers to the presumption of deep cultural unity of a citizenry, to the idea that representatives and represented are bound together in a tight circle of delegation, or to a single model of democratic rule. Rather, the value of democracy inheres in the quality of *relations* between different imperatives, interests, and identities – that is, it lies in the degree to which definitions of the proper balance between imperatives of collective action and individual freedom, between conflicting interests, and between multiple and fluid identities remain open to contestation and challenge.

Spaces of Democracy

We have suggested that the universalization of democracy as a taken-for-granted good does not imply that the meaning of democracy is cut and dried. Quite the contrary, it has coincided with a flowering of critical accounts of democratic theory and practice. If, at a minimum, this universalization indicates that there is no alternative to the legitimization of rule by reference to the will of the people, then it also indicates the point at which the elusive qualities of 'the people' become all the more evident (Offe, 1996). The questions of just *who* should participate, *how* this participation is going to be arranged, and *what* scope of actions are to be subjected to democratic oversight, have become more problematic, not less, with the historical 'triumph' of democratic norms. It is these three dimensions – the who, how, and what of democracy – that the chapters in this book address. They all share a strong commitment that the geographies of democracy are deeply implicated in working out practical solutions to these questions of democracy's meaning. Each chapter sets out to connect the practicalities of democracy with questions of democratic theory, without idealizing democracy or collapsing normative reflection into a priori models of desirable end-states. Taken together, they underscore the need to explore democracy as a specific sort of politics that constantly invites the evaluation and appraisal of first principles.

We have divided the chapters into three broad sections. The first section, *Elections, Voting and Representation*, addresses the complex and changing meanings of some of the basic mechanisms of modern democracy. The opening chapter addresses the basic context for the whole collection, namely the geographies of democracy's diffusion. John O'Loughlin provides a critical evaluation of the empirical and conceptual assumptions that inform the measurement and evaluation of democratization processes among academics, policy-makers, think-tanks, and politicians. The next two chapters, by Ron Johnston and Charles Pattie on the uneasy relationship between electoral geography and political science, and by Richard Morrill on the politics of electoral re-districting in the United States, both develop critical insights into perhaps the basic institutions of modern democracy – elections. Taken together, these two chapters illustrate the complexity of representative and representational practices involved in the design, implementation, and interpretation of democratic electoral politics.

The second section, ***Democracy, Citizenship and Scale***, raises questions concerning the spaces within which democratic politics takes place, and in particular the relations between different spaces of democracy – between domestic spaces and national polities, between the spaces of cities and wider regional and national scales, and between national-level politics and international processes of migration. The three chapters each explore the implications of thinking seriously about the complex spatialities and the constructed scales of democratic polities. Sallie Marston and Katharyne Mitchell develop a critical account of the changing geographies of citizenship. They illustrate the variability of citizenship identities and practices in relation to scales of local state, domestic space, the nation and, increasingly, transnational networks of migration. Their key contribution is the notion of citizenship-formation, calling attention to the institutions, social relations, and embodied practices through which the meaning of citizenship is made up and transformed in different contexts. David Smith addresses a fundamental tension within liberal theories of democratic legitimacy, namely whether there are any legitimate grounds to exclude outsiders from full citizenship status. At stake in his discussion is the fundamental question of the scope of the basic unit of democratic theory itself, the political community. There has been a great deal of discussion recently over whether globalization spells the death-knell of national democracy, suggesting that democracy's 'real' level is lower down, at the scale of the region, the locality, or the city. Murray Low explores the limitations of these arguments by examining the relationships of dependence and interdependence between democracy at sub-national scales and national level decision-making.

The final section, ***Making Democratic Spaces***, considers the identity and location of a broad range of informal types of politics, which are essential to the vibrancy of democracy and democratization. It includes considerations of the concept of public space, the importance of cultural practice in underwriting robust democratic public life, and the changing role of social movements in a globalizing world. The first three chapters in this section address another central conundrum of democratic theory, namely the identity and location of the collective subject of democratic politics, the public. Lynn Staeheli and Don Mitchell explore the changing meanings of the public/private distinction. They suggest that public action can take place in putatively private spaces, but also that what are nominally public spaces are increasingly subjected to processes of exclusionary privatization. Gareth Jones develops similar themes, exploring the practices and performances through which new forms of public space have been developed and sustained in the context of democratization in Latin America. The strong emphasis of his analysis is upon public space as a realm of communication between different social subjects. This theme is further developed in Clive Barnett's chapter. He argues against overly concrete conceptions of public space and overly substantive conceptions of the public, suggesting instead that stretched-out, mediated forms of communication be thought of as the space of democratic politics.

These three chapters all touch on the cultural infrastructure that underpins democratic politics, and that sustains practices of tolerance, respect, and acknowledgement. This theme is further developed in the following chapter by Sophie Watson, who argues that Robert Putnam's influential account of the relationship between social capital and the quality of democratic governance clings to a narrow understanding of the forms of cultural and social interaction that sustain a democratic ethos. She suggests that this approach, with its in-built tendency to see only decline in the trajectory of contemporary social trends, is looking in the wrong places for signs of vibrant democratic cultures, and in turn, looking at the wrong people – ignoring the emergent democratic subjectivities of organized women's groups, youth cultures, and the elderly, among others. Finally, and developing the emphasis in previous chapters on the importance of citizen action and cultural practices in democratization processes, Byron Miller picks up one of the most pressing questions of contemporary democratic politics – the role and future of social movement mobilization as a force for establishing, sustaining and deepening democracy. Miller's discussion focuses in particular on the challenge of globalization for both the conceptualization and the practice of social movement mobilization, and critically assesses the possibilities and limitations of emergent forms of transnational movement mobilizations.

In line with the preceding discussion in this Introduction, the combination of chapters in this book therefore aims to do two things. On the one hand, the chapters address a broad range of arenas and actors through which the scope and meaning of democracy has been extended and deepened, including the media, social movements, community mobilization, and patterns of associational culture. At the same time, they open up new questions about some well-established fields of state-centred democratic politics, reconsidering the nature of elections and electoral systems, central–local state relations, and political membership. We hope that, in bringing leading-edge theorizations of space, place, and scale to bear on existing conceptualizations of democracy, the collection will put normative questions of democracy, justice and legitimacy at the centre of critical geographic analysis of contemporary socio-economic transformations.

References

Agnew, J. (1994). 'The territorial trap: the geographical assumptions of international relations theory', *Review of International Political Economy*, 1: 53–80.
Agnew, J. (1998). *Geopolitics*. London: Routledge.
Agnew, J. (2002). *Making Political Geography*. London: Arnold.
Allen, J. (2003). *Lost Geographies of Power*. London: Blackwell.
Amin, A. (2002). 'Spatialities of globalisation', *Environment and Planning A*, 34: 385–99.
Amin, A. and Thrift, N. (2002). *Cities*, Cambridge: Polity Press.
Anderson, J. (2002). *Transnational Democracy: Political Spaces and Border Crossings*. London: Routledge.
Barnett, C. (2004). 'A critique of the cultural turn', in J. Duncan et al. (eds), *A Companion to Cultural Geography*. Oxford: Blackwell.
Benhabib, S. (1992). *Situating the Self*. Cambridge: Polity Press.
Benhabib, S. (2002). *The Claims of Culture: Equality and Diversity in the Global Era*. Princeton, NJ: Princeton University Press.
Brown, M. (1997). *Replacing Citizenship: AIDS Activism and Radical Democracy*. New York: Guilford Press.
Burgess, J., Harrison, C. and Filius, P. (1998). 'Environmental communication and the cultural politics of environmental citizenship', *Environment and Planning A*, 30: 1445–60.
Cole, P. (2000). *Philosophies of Exclusion: Liberal Political Theory and Immigration*. Edinburgh: Edinburgh University Press.
Connolly, W. (1991). 'Democracy and territoriality', *Millennium*, 20: 463–84.
Connolly, W. (1993). *The Terms of Political Discourse*. Oxford: Blackwell.
Connolly, W. (2000). 'Speed, concentric cultures, and cosmopolitanism', *Political Theory*, 29: 596–618.
Corbridge, S. (1993). 'Marxisms, modernities and moralities: development praxis and the claims of distant strangers', *Environment and Planning D: Society and Space*, 11: 449–72.
Corbridge, S. (1998). 'Reading David Harvey: entries, voices, loyalties', *Antipode*, 30: 43–55.
Dahl, R. (1989). *Democracy and Its Critics*. New Haven, CT: Yale University Press.
Dahl, R. and Tufte, E. (1973). *Size and Democracy*. Stanford, CA: Stanford University Press.
Derrida, J. (1992). *The Other Heading*. Bloomington: Indiana University Press.
Derrida, J. (1997). *The Politics of Friendship*. London: Verso.
Derrida, J. (2001). *On Cosmopolitanism and Forgiveness*. London: Routledge.
Dietz, M. (1998). 'Merely combating the phrases of this world', *Political Theory*, 26: 112–39.
Dunn, J. (1992). 'Conclusion', in J. Dunn (ed.), *Democracy: The Unfinished Journey, 508 BC to AD 1993*. Oxford: Oxford University Press, pp. 239–66.
Dunn, J. (1999) 'Situating democratic political accountability', in A. Przeworski et al. (eds), *Democracy, Accountability and Representation*. Cambridge: Cambridge University Press, pp. 329–44.
Entrikin, N. (1999). 'Political community, identity and cosmopolitan place', *International Sociology*, 14: 269–82.

Fukuyama, F. (1993). *The End of History and the Last Man*. Harmondsworth: Penguin.

Gallie, W. B. (1956). 'Essentially contested concepts', *Proceedings of the Aristotelian Society*, 56: 167–98.

Gaus, G. (2003) *Contemporary Theories of Liberalism*. London: Sage.

Habermas, J. (1996). *Between Facts and Norms*. Cambridge: Polity Press.

Habermas, J. (1998). *The Inclusion of the Other*. Cambridge: Polity Press.

Hajer, M. and Kesselring, S. (1999). 'Democracy in the risk society? Learning from the new politics of mobility in Munich', *Environmental Politics*, 8(3): 1–23.

Hanagan, M. and Tilly, C. (eds) (1999). *Extending Citizenship, Reconfiguring State*. Lanham, MD: Rowman and Littlefield.

Hannah, M. (2001). 'Sampling and the politics of representation in US Census 2000', *Environment and Planning D: Society and Space*, 19: 515–34.

Harvey, D. (2000). 'Cosmopolitanism and the banality of geographical evils', *Public Culture*, 12 (2): 529–64.

Held, D. (1997). 'Democracy: from city-states to a cosmopolitan order?', in R. E. Goodin and P. Pettit (eds), *Contemporary Political Philosophy*. Oxford: Blackwell, pp. 78–101.

Honig, B. (2001). *Democracy and the Foreigner*. Princeton, NJ: Princeton University Press.

Howell, P. (1993). 'Public space and the public sphere: political theory and the historical geography of modernity', *Environment and Planning D: Society and Space*, 11: 303–22.

Jeffrey, C. J. (2000). 'Democratisation without representation? The power and political strategies of a rural elite in north India', *Political Geography*, 19: 1013–36.

Johnston, R. J. (1999). 'Geography, fairness, and liberal democracy', in J. Proctor and D. M. Smith (eds), *Geography and Ethics*. London: Routledge, pp. 44–58.

Jones, J. P. and Moss, P. (1995). 'Democracy, identity and space', *Environment and Planning D: Society and Space*, 13.

Katznelson, I. (1996). 'Social justice, liberalism and the city', in A. Merrifield and E. Swyngedouw (eds), *The Urbanization of Injustice*. London: Lawrence and Wishart, pp. 45–64.

Laclau, E. and Mouffe, C. (1985). *Hegemony and Socialist Strategy*. London: Verso.

Lefort, C. (2000). *Writing: The Political Test*. Durham, NC: Duke University Press.

Low, M. (1997). 'Representation unbound', in K. Cox (ed.), *Spaces of Globalization*. New York: Guilford.

Low, M. (2000). 'States, legitimacy and collective action', in Peter Daniels et al. (eds), *Human Geography: Challenges for the 21st Century*. London: Longman, pp. 473–502.

Low, M. (2003). 'Political geography in question', *Political Geography*, 22: 625–31.

Manin, B. (1997). *The Principles of Representative Government*. Cambridge: Cambridge University Press.

Mason, M. (2001). 'Transnational environmental obligations: locating new spaces of accountability in a post-Westphalian global order', *Transactions of the Institute of British Geographers*, 26: 407–29.

Mercer, C. (2002). 'NGOs, civil society and democratisation', *Progress in Development Studies*, 2: 5–22.

McIlwaine, C. (1998). 'Civil society and development geography', *Progress in Human Geography*, 22: 415–24.

Mouffe, C. (1998). *The Democratic Paradox*. London: Verso.

Nancy, J.-L. (1991). *The Inoperative Community*. Minneapolis: University of Minnesota.

Nussbaum, M. (2000). *Women and Human Development: The Capabilities Approach*. Cambridge: Cambridge University Press.

Offe, C. (1996). 'Constitutional policy in search of the "Will of the People"', in *Modernity and the State*. Cambridge: Polity Press.

O'Neill, J. (2001). 'Representing people, representing nature, representing the world', *Environment and Planning C: Government and Policy*, 19: 483–500.

O'Neill, O. (2000). *The Bounds of Justice*. Cambridge: Cambridge University Press.

Owens, S. (2001). 'Engaging the public: information and deliberation in environmental policy', *Environment and Planning A*, 32: 1141–8

Parekh, B. (1993). 'The cultural particularity of liberal democracy', in D. Held (ed.), *Prospects for Democracy*. Cambridge: Polity Press, pp. 156–75.

Phillips, A. (1991). *Engendering Democracy*. Cambridge: Polity Press.

Rawls, J. (1971). *A Theory of Justice*. Cambridge, MA: Havard University Press.

Proctor, J. and Smith, D. M. (1999). *Geography and Ethics*. London: Routledge.

Robinson, J. (1998). 'Spaces of democracy: re-mapping the apartheid city', *Environment and Planning D: Society and Space*, 16: 533–48.

Sayer, A. and Storper, M. (1997). 'Ethics unbound: for a normative turn in social theory', *Environment and Planning D: Society and Space*, 15: 1–17.

Schaffer, R. (1998). *Democracy in Translation: Understanding Politics in an Unfamiliar Culture*. Ithaca, NY: Cornell University Press.

Schmitter, P. (1999). 'The future of democracy: could it be a matter of scale?', *Social Research*, 66: 933–58.

Sen, A. (1999a). 'Democracy as a universal value', *Journal of Democracy*, 10 (3): 3–17.

Sen, A. (1999b). *Development as Freedom*. Oxford: Oxford University Press.

Sewell, W. H. (2001). 'Space in contentious politics', in J. Aminzade et al. (eds), *Silence and Voice in the Study of Contentious Politics*. Cambridge: Cambridge University Press.

Shapiro, I. (1999). *Democratic Justice*. New Haven, CT: Yale University Press.

Shapiro, I. and Hacker-Cordón, C. (eds) (1999). *Democracy's Edges*. Cambridge: Cambridge University Press.

Sharp, J., Routledge, P., Philo, C. and Paddison, R. (eds) (1999). *Entanglements of Power: Geographies of Domination/Resistance*. London: Routledge.

Slater, D. (2002). 'Other domains of democratic theory: space, power, and the politics of democratization', *Environment and Planning D: Society and Space*, 20: 255–76.

Spinosa, C., Flores, F. and Dreyfus, H. (1999). *Disclosing New Worlds: Entrepreneurship, Democratic Action, and the Cultivation of Solidarity*. Cambridge, MA: MIT Press.

Squires, J. (2002). 'Democracy as flawed hegemon', *Economy and Society*, 31: 132–51.

Taylor, P. (1994). 'The state as container: territoriality in the modern world-system', *Progress in Human Geography*, 18: 151–62.

Taylor, P. (1995). 'Beyond containers: internationality, interstateness, and interterritoriality', *Progress in Human Geography*, 19: 1–15.

Walker, R. B. J. (1993). *Inside/Outside: International Relations as Political Theory*. Cambridge. Cambridge University Press.

White, S. (2000). *Sustaining Affirmation: The Strengths of Weak Ontology*. Princeton, NJ: Princeton University Press.

Williams, G., Veron, R., Corbridge, S. and Srivastava, M. (2003). 'Participation and power: poor peoples engagement with India's employment assurance scheme', *Development and Change*, 34: 163–92.

Young, I. M. (1990). *Justice and the Politics of Difference*, Princeton, NJ: Princeton University Press.

Young, I. M. (1993). 'Justice and communicative democracy', in R. Gottleib (ed.), *Radical Philosophy*. Philadelphia: Temple University Press, pp. 123–43.

Young, I. M. (1998). 'Harvey's complaint with race and gender struggles: a critical response', *Antipode*, 30: 36–42.

2 Global Democratization: Measuring and Explaining the Diffusion of Democracy

John O'Loughlin

Since Francis Fukuyama (1992: xi) declared that the 'End of History' had been reached because liberal democracy constitutes the 'endpoint of mankind's ideological motivation' and is 'the final form of human government', a parallel debate has raged about whether liberal democracy, as practiced in the West, will diffuse and be accepted throughout the rest of the world. By the turn of the twenty-first century, few political leaders – even in authoritarian states – were willing to argue aloud against democracy since its virtues are now almost universally accepted. Global norms are fast coalescing around some key human and political freedoms, starting with the Universal Declaration of Human Rights (adopted by the UN General Assembly in 1948) and extending to the 1993 World Conference on Human Rights in Vienna (Diamond, 1999: 4). Democracy is essential to freedom and other inalienable rights because (a) free and fair elections require certain political rights of expression and these will co-exist with other liberties; (b) democracy maximises the opportunity for self-determination; and (c) democracy facilitates moral autonomy, the ability to make normative choices and to be self-governing (Diamond, 1999).

Nobody disputes that the number of democratic states rose dramatically after the collapse of the Soviet Union and the end of Communist regimes in Eastern Europe. What remains in question is whether the new democracies are (a) stable; (b) truly democratic or only veneer expressions of democracy while real power still rests with autocrats; and (c) whether there is a general global process or whether recent developments are the independent results of separate and unpredictable domestic circumstances. After the end of the Cold War, a paradigm shift is recognisable in the study of democratisation. Rather than seeing political developments as separate events, researchers turned to seeing them as connected within a cascading pattern and thus part of a 'Third Wave' of democratisation (Huntington, 1991). The structural model of predictability implicit in a cascading wave, in turn presupposing a structural trend, is now viewed skeptically by students of comparative politics who focus on national differences (Schwartzman, 1998). In political science, an argument has erupted about whether one can compare polities across regions (Inglehart and Carballo, 1997).

Geographers and some political scientists (Most and Starr, 1989; Siverson and Starr, 1991) reject the binary choice of a particularist versus a structuralist perspective on global political change. Instead, a 'domain-specific model' is preferred in which both general global trends and local circumstances are examined in an interactive manner. In statistical terms, it means the fitting of a predictive regression model that uses the characteristics of the states to anticipate political changes and it specifically identifies those countries that do not conform to the general trends to highlight what makes them different. O'Loughlin et al. (1998) used a diffusion model to track the democratic and autocratic changes after 1946 but they were highly cognisant of both regional peculiarities and states that did not conform to the regional trends. In this chapter, I extend this perspective. Further evidence for the efficacy of the diffusion model of democracy suggests that it offers a vibrant option that can incorporate the special contexts of individual countries and the predictability inherent in the general model of global change (O'Loughlin, 2001). The 'context-specific' approach of geographers has been widely applauded within the discipline (Agnew, 1996; Dorling, 2001) but viewed skeptically by some political scientists (King, 1996) who think that it represents a missing variables problem. Some key explanatory variables for the patterns are not considered by the geographers who, like their comparative politics counterparts, are intent on promoting a place-specific approach (O'Loughlin, 2000).

In this chapter, I take stock of the democratisation trends since the mid-1990s. While it appeared for about half a decade after the collapse of the Communist regimes that the world was firmly ensconced in the 'Third Wave' of democratisation (the first two were in the nineteenth century and after World War I, followed by reversals to authoritarianism in both cases), recent evidence is more contradictory. The reversal to authoritarianism that was anticipated by Huntington's account of the 'Third Wave' of democratisation of the late 1970s, 1980s and early 1990s has not yet happened in a dramatic manner, but neither has the 'wave' continued its upward trajectory. Instead, the beginning of the twenty-first century marks a period of stability in the democratic trend. As noted by Norris (1999: 265), the percentage of independent states that were democratic (according to the Freedom House data on political and civil rights) was 34% in 1983 and rose to 41% in 1997, where it has remained. What is especially noticeable about the trend in the 1990s is the strong macro-regional character of the democratic transitions – it is clear regional location matters.

The specific purpose of this chapter is to probe the causes of the turn to 'democratisation', without prejudging whether the wave is real or imaginary. There are five issues that need to be considered. First, is there a new international norm consequent on globalisation? Is there a political parallel to economic globalisation that is making countries politically similar? (Held et al., 1999)? Second, is there a clear correlation between democracy and aid: no democracy, no foreign aid? There is little doubt that democracy has been strongly promoted by the United States and its allies, and that economic development strategies by international and national agencies are intimately linked to grassroots democracy initiatives and transparent governance. Third, is the diffusion effect identified at the time of the end of the Cold War still evident, or has the asymptote[1] been reached? Exceptions, both in regional and local terms, to general diffusion trends can be especially instructive in suggesting future trends. Fourth, with the collapse of the Communist alternative about 1989–91 and the sweep of the democratic idea worldwide, can we accept the 'Zeitgeist' model from Linz and Stepan (1996: 74): 'When a country is part of an international ideological community where democracy is only one of many strongly contested ideologies, the chances of transitioning to and consolidating democracy are substantially less than if the spirit of the times is one where democratic ideologies have no powerful contenders.' If the 'Zeitgeist' exerts such a strong control, then we would not expect a Third Wave of reversals as happened after the two earlier waves of democratisation. Finally, the end-game of 'Zeitgeist' democracy management is the development of a cosmopolitan political culture worldwide, one that is not only promoted, but whose causes are multiple and indefinite. Archibughu et al. (1998), Held (1993) and Risse et al. (1999) have developed the concept of a 'cosmopolitan political culture'. Lynch (2001) considers the tension between, on the one hand, state domestic political cultures and, on the other hand, the international governance advocated by Richard Falk (2000), who wants to strengthen and develop global political institutions. The cosmopolitans want to go further than simply promote democracy, though that is clearly a first step in their project. They believe that the state system is manifestly inefficient and that a global political arena can replace the conditions and the dynamics of both domestic and international politics without the corresponding emergence of an international state. In the extreme version, the development of an 'international domestic politics' of democratically legitimated decisions consequent on the emergence of a globalised political arena is envisaged (Lynch, 2001: 93).

Democracy: Transitions and Measurement

Contemporary research on the distribution of democracy was kicked off by Lipset's (1959) paper on the social requisites of democracy. This focused attention on the structural characteristics of countries, typically the size of the middle class, private entrepreneurial groups, widespread literacy, and sustaining civic values. Recent updates of this approach have typically been able to replicate the original conclusions of the 'social requisites' school (Lipset et al., 1993), though attention now has been diverted somewhat to issues of democratic reversal, democratic consolidation, and democratic transitions. In a paper responding to Lipset, Rustow (1970) argued that the structural national conditions that keep a democracy functioning might not be the same factors that brought democracy to the country in the first place. Focusing on 'contingent conditions' and dynamic circumstances, he deviated from the Lipset position since he believed it too narrow and limiting. Rustow offered an alternative of 'a more varied mix of economic and cultural dispositions with contingent developments and individual choices' (Anderson, 1999: 2).

Democratic transitions typically occur in stages and are the by-products of debate, struggle, compromise, and agreement. It has been shown empirically that a state that has a chance of becoming democratic will have a sense of community, a conscious adoption of democratic rules, and operation of the rules in a step-by-step adoption of democracy (Rustow, 1970). Political elites are the key actors, whether in government or opposition, and elite-bargaining is an element of all transitions (Bermeo, 1999; Haggard and Kaufman, 1999). Democratic transitions are especially tenuous in times of economic uncertainty since economic decline can reverse democratic trends by giving rise to social unrest and class strife. In low-income countries, democracy has a 12% chance of breakdown in any given year (sampled between 1950 and 1990) and the expected life of a democracy increases with per capita income. Further, democracy is more likely to endure when income inequality is lower (Przeworski, 1991).

Globalisation is seen as a 'global catalyst' of democracy by Schwartzman (1998) from her survey of the democratisation literature. Global industrialisation and development filter through to democracy in four ways: (1) they privilege the role of technology and communication, making the import of ideas easier and therefore more difficult for an authoritarian regime to control; (2) they promote the growth of the middle class in individual countries, a key factor in the pressure for democratisation; (3) they increase the power of the working class, a key pressure group according to Rueschemeyer et al. (1992); and (4) they exaggerate the interplay between globalised capitalism and state–class relations, thus producing domestic pressures for political change. But globalisation involves more than economic linkages. Additional external influences are transmitted by new technologies, including the Internet, satellite dishes, international TV networks, and instant news dissemination:

> Communications technology has reshaped the opportunity structure of contemporary politics, making almost every political issue one of international rather than purely domestic interest. Many of the political manifestations of globalization such as the rise of intrusive human rights norms and the proliferation of international and transnational organisations, can plausibly be accounted for within the context of communications technology. (Lynch, 2001: 95)

While it is common to assume that the positive relationship between globalisation and democracy is becoming stronger, Moon (1996: 10) cautions that 'democracy and globalisation have not been necessarily complementary. They have often produced ambiguous and conflicting implications.'

But what is democracy? How should it be measured? More than 550 sub-types of democracy are identified in Collier and Levitsky's (1997) review of 150 studies. Minimalist definitions of democracy derive from Joseph Schumpeter (1947: 269), who defined democracy as a system 'for arriving at political decisions in which individuals acquire the power to decide by means of a competitive struggle for the peoples' vote'. This minimalist approach was adhered to by Huntington (1991). Dahl's (1971) concept of polyarchy has two overt dimensions: *opposition* (organised contestation through regular, free and fair elections); and *participation* (the right of virtually every adult to vote and contest for office). Embedded in these two notions is a third concept: civil liberty (Diamond, 1999: 8).

The minimalist definition of democracy corresponds to *formal democracy* with four common features: regular fair and free elections; universal suffrage; accountability of the state's apparatus to the people; and effective guarantees of expression and association (Beetham, 1994). Adding another condition, high levels of democratic participation without systematic differences across social categories makes for *participatory democracy*. A key difference between those who study formal democracy and those who equate democracy to popular democracy is that the latter want social and economic equality as well (Bobbio, 1989). Yet another condition, increasing equality in social and economic outcomes, produces *social democracy* (Huber et al., 1999). In Huber et al.'s model, formal democracy opens the door for the other democratic forms and a virtuous cycle of egalitarian policies and norms allow more citizens to participate in the political process. However, this pattern of development is far from automatic. In Latin America, for example, formal democracy developed only partially. External pressures, especially from the United States, favour a deepening of formal democracy but typically block implementation of principles and practices that promote participatory and social democracy (Huber et al., 1999; see also, Boeninger, 1997; Bollen, 1993; Rueschemeyer et al., 1992).

Formal democracy should also be distinguished from *liberal democracy* – formal democracy that encompasses extensive protections for individual and group freedoms, inclusive pluralism in civil society and party politics, and civilian control over the military. Using the Freedom House scores (see below for more details on these measures), only 41.4% of countries were 'free' (liberal democratic in character) in 1996, though the percentage of formal democracies was just over 60% of the world's states. This sizeable difference led Diamond (1997: xv) to conclude that the 'Third Wave' of democratisation has had much greater breadth than depth, and that outside the wealthy industrialised countries, liberal democracy tends to be shallow, illiberal and poorly institutionalised (Zakaria, 1997). Three features distinguish liberal democracies from electoral democracies: (1) an absence of domains of power for the military and others not accountable to the electorate; (2) the requirement of horizontal accountability that office holders owe to each other; and (3) extensive provisions for political and civic pluralism as well as personal and group freedoms (Diamond, 1999: 8). For example, Turkey, Ukraine, Georgia, Zambia and Russia are not (yet) liberal democracies, since political violence, lawlessness and corruption are still a significant feature in these states. While traditionally associated with Latin America in democratisation research, hollow, illiberal, poorly institutionalised democracy is by no means unique to that region and is now characteristic of many Third Wave democracies. These 'pseudo-democracies' have regular elections and political parties. Electoral outcomes are uncertain because the competition is real between the cadres of the elites. However, mass parties and grassroots democratic movements are noticeably absent from the political scene.

Data from 1946–94, using a measure for democracy based on authority characteristics, show that democratisation has proceeded in regular spatial and temporal diffusion patterns, but with distinct observable regional trends (O'Loughlin et al., 1998). Unlike the trend suggested by Huntington's Third Wave model, this suggests a more complex process. Regional-level explanations, rather than macro-structural ones are necessary to account for the political changes of the past half-century. The geographic disparities in the global trends in democratisation had barely been mentioned in previous global-level analyses. This raises the question of whether the geographic factor was simply an artifact of an approach that emphasised the 'spatial and temporal diffusion of democracy'? Or was it a result of the special combination of place characteristics that mold a certain style of politics, and that cannot easily be isolated from socio-demographic explanations to which other social scientists resort?

It would not be going too far to claim that democracy's meaning is to some extent place-specific and that, global trends notwithstanding, sharp differences between places are evident even within the set of stable democratic countries. Any world map of the distribution of democratic scores indicates clear regional clusters and temporal framing: the past 50 years also indicates the regional ebb and flow of democracy in a distinctly time–space autocorrelation (O'Loughlin et al., 1998). What is less evident is the combination of forces generating these clusters in time and space. Some sort of regional neighborhood-effect is plausible, especially in sub-Saharan African and Latin American countries in the early 1990s. Of course, the best-known regional trend was in Eastern Europe after 1980 but that trend did not reach all of the former Communist states; Central Asia, to name one region, remains markedly different than Central Europe in its democratic qualities today. As Kopstein and Reilly (2000) conclude, geography seems to be more important than democratic policies, as the contrasting examples of Kyrgyzstan (many policy reforms but less democratic) and Slovakia (fewer democratic policies but more democracy) demonstrate.

The diffusion-promotion effects of a neo-liberal world order put in place since the collapse of the Soviet Union in the late 1980s and the specific regional interests of the powerful Western countries, especially the United States, are instrumental in establishing the external push factors, giving the world a peculiar dynamic at the present time (Joseph, 1997). Contemporary democratisation requires concessions from those who were formerly excluded from participation. Having tolerated many years of material inequities while at the same time agreeing to work through elections and democratic procedures, these dissenters must now wait longer. However, they generally have the pressures of international agencies and benefactors on their side. Regardless of the academic debate about the democratic trend, it is clear that democracy and international politics are now intertwined for Western countries. Former US Under-Secretary of State Strobe Talbott (1996: 63) put it starkly: 'Only in an increasingly democratic world will the American people feel themselves truly secure.'

Analyzing Democratic Diffusion: The Freedom House Measures

As noted above, the choice of democratic measure is not as evident as it might seem at first glance. Of the myriad of indicators that are now readily available, the preferred one should be able to summarise more than one element of the global democratic profile. Because electoral democratic measures, such as the Polity III[2] measures used by O'Loughlin et al. (1998), are limited to formal democratic institutions, a fuller picture of democracy needs to consider other, less institutionalised measures. Here, I use the Freedom House[3] scores that incorporate the concepts of liberal democracy for my analysis. Both Freedom House and Human Rights Watch,[4] though different in ideological orientation, worry about the growing gap between formal and liberal democracies. Though accused in the United Nations by authoritarian states such as Cuba, China and Sudan of being biased, the Freedom House measures of political and civil rights have been used widely in academic work, not least because they have been available for 30 years.

Freedom House carries out a yearly survey of all countries. The survey rates countries and territories by focusing on the rights and freedoms enjoyed by individuals in each country or territory: 'To reach its conclusions, the survey team employs a broad range of international sources of information, including both foreign and domestic news reports, NGO publications, think tank and academic analyses, and individual professional contacts. The survey's understanding of freedom encompasses two general sets of characteristics grouped under political rights and civil liberties. Political rights enable people to participate freely in the political process, which is the system by which the polity chooses authoritative policy makers and attempts to make binding decisions affecting the national, regional, or local community. In a free society, this represents the right of all adults to vote and compete for public office, and for elected representatives to have a decisive vote on public policies. Civil liberties include the freedoms to develop views, institutions, and personal autonomy apart from the state. The survey employs two series of checklists, one for questions regarding political rights and one for civil liberties, and assigns each country or territory considered a numerical rating for each category (Freedom House, 2000). In the Freedom

House scores, 1 indicates the highest ranking (freedom on the political rights dimension) of democracy, with 7 indicating the most authoritarian regimes. Similarly, 1 to 7 on the civil liberties scale ranks freedom of beliefs and expression. The 'state of freedom' is gauged by Freedom House by assigning each country the status of 'free', 'partly free', or 'not free' through averaging their political rights and civil liberties ratings. Those whose ratings average 1–2.5 are generally considered 'free', 3–5.5 'partly free', and 5.5–7 'not free'. The dividing line between 'partly free' and 'not free' usually falls within the group whose ratings numbers average 5.5.[5]

In order to examine the global democratisation trends, I investigated the distributions of political rights and civil liberties in 1979 and 2001 and the changes between these two years. Though the Freedom House scores reach back to 1972 for many countries, many states and territories are missing data for the early years of the survey. The 22-year gap from 1979 to 2001 allows an adequate picture of the developments over the past quarter-century since the beginning of Huntington's Third Wave of democracy. The dramatic growth of two elements of liberal democracy, political rights and civil liberties, over the past two decades is evident in Figure 2.1. On the two graphs, a negative value indicates an improvement in these indices (low values reflect more democracy). It is evident that the overwhelming change between 1979 and 2001 is towards more democracy globally, with over twice as many countries becoming more democratic than have become less democratic – on both indices. There is, of course, a strong correlation between political rights and civil liberties across all countries, though a nuanced analysis of the two graphs indicates that slightly more countries have increased their political rights scores. A couple of outliers on the political rights graph (changes of +4 and +5) are markedly at odds with the global trends, while such dramatic developments are not as visible on the civil rights chart. About one-third of all countries did not change their scores across the two decades as stable democracies were more evident and many countries classified as 'partly free' by Freedom House retained their respective status.

A clearer picture of global democratisation can be obtained from Figures 2.2 and 2.3. For ease of interpretation, the maps have been simplified somewhat so that the map categories can be considered as 'more democratic', 'slight change towards democracy' (– 1), 'no change', and 'less democratic' (positive scores). Most of the stable scores are for democratic states in Western Europe, North America, etc., though a few authoritarian countries hung on to that status in the face of a global trend. Algeria, Mauritania, Syria, North Korea, Zimbabwe, Laos, Vietnam, Myanmar and Afghanistan are emblematic of the hold-outs from the global democratic wave. Other countries that were 'partly democratic' in 1979, such as Peru, Guatemala, Oman, Zambia and Pakistan, held on to that categorisation. By contrast, the clearest expression of global democratisation lies in the previously Communist countries of Eastern Europe and the former Soviet Union. The changes in these areas were not uniform, however. While countries such as Mongolia, Poland, Bulgaria, and the Baltic Republics showed dramatic

Political Rights 1979–2001

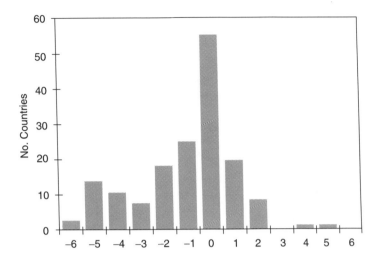

Civil Liberties 1979–2001

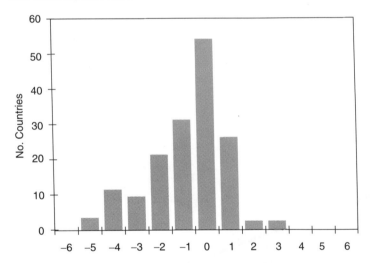

Figure 2.1   Changes in political and civil rights 1979–2001

gains, other former Soviet republics such as Belarus, Azerbaijan, Tajikistan and Kyrgyzstan replaced one kind of political authoritarianism for another. Worse yet, some former Soviet republics are scored as more repressive than in the last decade of the Soviet Union. Three of the five Central Asian republics (Kazakhstan, Turkmenistan and Uzbekistan) as well as other Islamic states of the neighboring Middle East (Iran, Syria, Saudi Arabia, Egypt, Sudan, Libya,

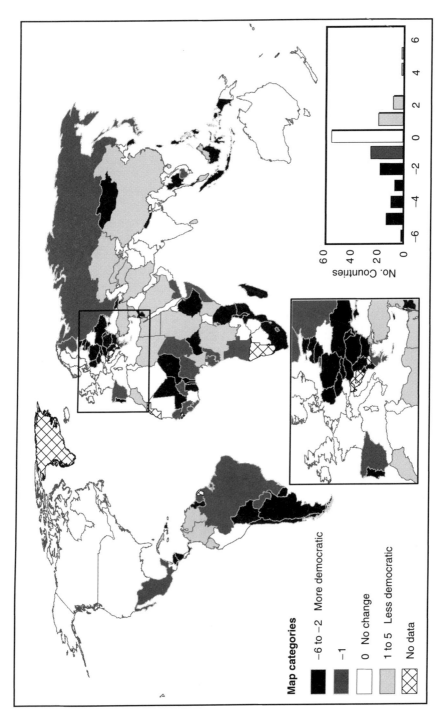

Figure 2.2 Changes in political rights 1979–2001

Map categories

■ −6 to −2 More democratic
▨ −1
□ 0 No change
▨ 1 to 5 Less democratic
▨ No data

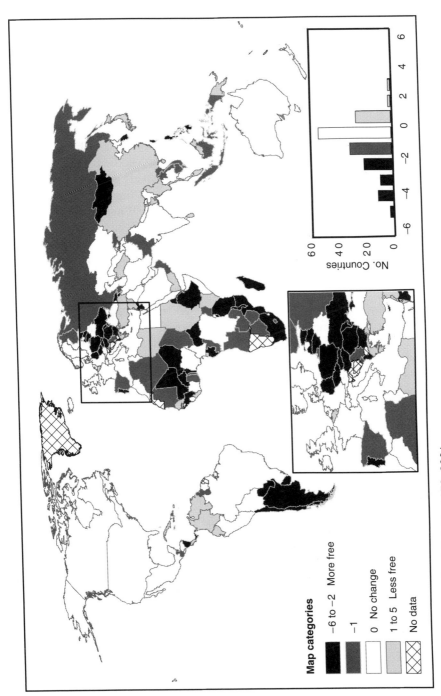

Figure 2.3 Changes in civil rights 1979–2001

Map categories

■ −6 to −2 More free
▨ −1
□ 0 No change
▨ 1 to 5 Less free
▧ No data

Morocco, Turkey) and China all saw a decline between 1979 and 2001 in political rights. While most of Latin America became more democratic, the northern Andean countries (Colombia, Venezuela and Ecuador) saw a reverse trend. Africa showed the greatest diversity in political rights; while most African states became more democratic, exemplified best by South Africa, three notable exceptions were the Democratic Republic of Congo, Kenya, and Gabon.

Since countries that value the political elements of liberal democracy tend to promote civil liberties as well, the close correlation between these two dimensions is visible in a comparison of Figures 2.2 and 2.3. Countries with a negative trend in their political rights score are also characterised by no improvement in their civil liberties index. Most Islamic states of North Africa, the Middle East and Central Asia have seen either no change or even reversal of civil liberty gains in recent years. By contrast, big improvements can be noted in the former Communist countries, in most of Africa, and in the southern cone of South America. While political and civil rights generally march in tandem, it is often the case that political developments, especially in the formation of new parties, precede the improvement in civil liberties. But the fact that the past two decades have seen more improvements in civil liberties than reversals should not generate a sense of inevitability. Reversals are common in democratising countries whose institutions are not stabilised and where grassroots support for democratic values is not yet widespread. The example of Turkey is instructive: after the military intervention in 1980, there was a sharp reversal in the treatment of the secular government's political opponents (especially Islamicists) and its largest minority group, the Kurds. More than any other indicator of liberal democracy, civil rights offers a deeper and more meaningful measure of democracy than the more accessible electoral measures of parliamentary competition. For that reason, more researchers use civil rights indicators: available sources include the annual reviews of every country from Amnesty International and the US Department of State, and documents from the UN Commissioner for Human Rights, as well as the Freedom House measures used here.

The state of play of contemporary democracy can be seen in the two maps (comprising Figures 2.4 and 2.5). The map categories follow the Freedom House nomenclature. On the political rights map, 50 countries are mapped as 'democratic' while another 39 are classed as 'mostly democratic'. Together, the other countries (partly and non-democratic) constitute about half of the world's polities and almost one-third are in the most repressive categories (non-democratic). Despite the gains of the Third Wave of democratisation, large regions of the globe are still relatively unaffected. A large swath from Central Africa through the Islamic world of North Africa, the Middle East and Central Asia to China and Indochina accounts for almost all non-democratic states in 2001. Previously non-democratic regions like South America, Southern Africa, Eastern Europe and the western half of the former Soviet Union are now at least partly democratic. Though reversals can be expected in some of these countries, the longer they remain in the democratic camp, the greater the likelihood of the establishment of an array of parliamentary elections and electoral turnover.

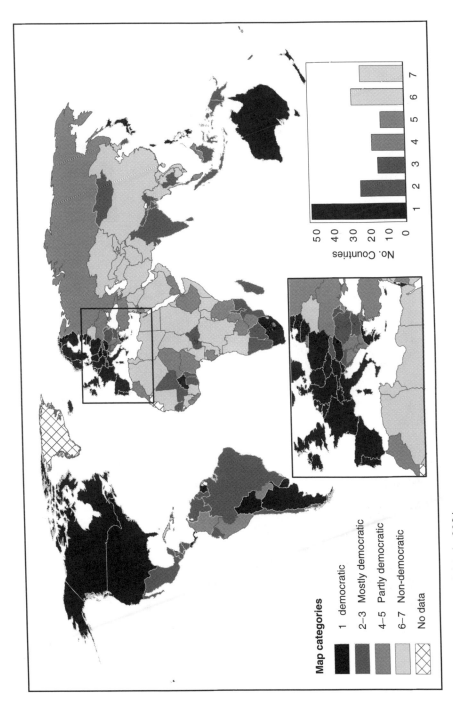

Figure 2.4   Political rights in 2001

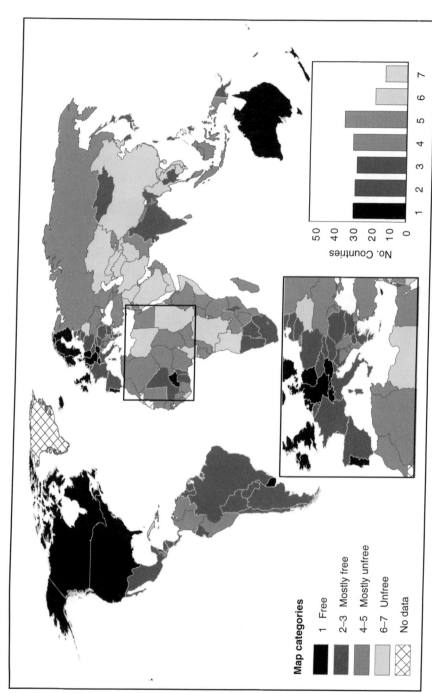

Figure 2.5 Civil rights in 2001

A comparison of the map of political rights to the final figure of civil liberties shows some clear, key differences that emerge in the expression of liberal democracy. Though countries that are 'partly democratic' are usually also 'mostly unfree', some differences emerge. Thus, although Pakistan, Egypt and Algeria are rated as 'non-democratic' for their lack of parliamentary democracy, they nonetheless have a modicum of civil rights ('mostly unfree'). The reverse, countries with lower ratings on civil liberties than political rights, include Poland, the Czech Republic, Hungary, Greece, Spain, and the Baltic Republics, all newly-democratic in the past quarter-century. Though there is little question of their democratic stability, their electoral and political procedures are more established than their implementation of other elements of liberal democracy in the form of civil liberties. Fair protection and treatment of minority groups (national, ethnic, religious, etc.) and political opponents is still the *sine qua non* of modern liberal democracy.

To explain the distributions on the maps, we must turn to the factors that are encouraging and promoting democratic diffusion. Some of these elements are necessarily idiosyncratic (internal factors in countries are most important) but others are more structural and connect to the notion of a democratic globalisation that has taken hold since the end of the Cold War. In particular, the key factors of contagious diffusion from neighboring states and promotion of democracy from the major Western countries must be examined to see how influential they have been in changing the world's political map.

Democracy: Diffusion and Promotion

Over the past 30 years, research on democracy has ebbed and flowed in its attention to the regional nature of political changes and regime characteristics. Lipset's (1959) paper on the social requisites of democracy focused attention on the structural characteristics of countries. Unlike the earlier version of the 'social requisites model', in a later paper Lipset (1994: 16) stated that 'a diffusion, a contagion or demonstration effect seems operative, as many have noted, one that encourages democracies to press for change and authoritarian rulers to give in' (see also Lipset et al., 1994). In his survey of the democratisation literature, Shin (1994: 153) concluded 'as vividly demonstrated in Eastern Europe and Latin America, earlier transitions to democracy have served as models for later transitions in other countries in the same region'. Huntington attributed the 'Third Wave' partly to a diffusion process starting in the Iberian peninsula in the mid-1970s:

> Successful democratisation occurs in one country, and this encourages democratisation in other countries, either because they seem to face similar problems, or because successful democratisation elsewhere suggests that democratisation might be a cure for their problems whatever their problems are, or because the country that has democratised is powerful and/or is viewed as a political and cultural model. (Huntington, 1991: 100)

As Huntington sees it, diffusion offers a proven course of action that can presumably be adopted and applied. It also works as a source of social learning by highlighting successes and failures. In his 'Third Wave', state leaders were able to observe clearly the processes as they unfolded, and could draw the obvious conclusion for their own domestic circumstances. But the diffusion effects were the strongest where proximity was the greatest and diffusion increased in effects over time. As O'Loughlin et al. (1998) show for sub-Saharan Africa and Latin America, the pattern of regime change over time shows a strong element of neighbor-to-neighbor linkages and also of cross-regional snowballing.

The 'peculiar dynamic' of the present time (Joseph, 1997) that is encouraging democratisation is having evident impacts. In sub-Saharan Africa in the 1990s, 'ruler conversions' to democratic forms and behavior took place quickly as leaders began to advise each other how to avoid being pressured into an unwelcome form of government (Joseph, 1997). What most worried these leaders was the potential loss of foreign aid; by the mid-1990s, Western governments were making it clear that they would withhold monies and assistance from authoritarian regimes. The United States Agency for International Development states the relationship between foreign aid and political objectives bluntly: 'US foreign assistance has always had the twofold purpose of furthering America's foreign policy interests in expanding democracy and free markets while improving the lives of the citizens of the developing world.'[6] The pressure applied to poor countries to democratise is part of a US-led strategy to build a more secure world order. As President Bill Clinton saw it in his second State of the Union message in 1995: 'Ultimately, the best strategy to ensure our security and to build a durable peace is to support the advance of democracy elsewhere [...]. The world's greatest democracy will lead a whole world of democracies.' According to the democratic peace hypothesis, democracies generally do not fight each other (Russett, 1993), and behave better towards their own citizens than less democratic states.

US policy, though voicing support for global democracy since the Presidency of Woodrow Wilson's commitment during World War I to 'save the world for democracy', has not always been consistent in tone, in strategy, in economic and political support, nor even in ideals. President Ronald Reagan devised the democracy crusade as an anti-Soviet policy but Presidents George Bush (Sr.) and Bill Clinton asserted that democracy promotion was a key organising principle of US foreign policy after the Cold War. What was a heightened moral dimension in the Cold War for Reagan was a strategy for peace in the post-Cold War world for his successors. A collection of articles published in the waning years of the Cold War advocated an even more 'evangelistic' mission for American democracy as a counterweight to the attractions of the Soviet model (Goldman and Douglas, 1988). But US policy has not been consistent, ignoring human rights and democracy in Kazakhstan, Saudi Arabia, Egypt, China, Indonesia before 1998, Armenia and Azerbaijan but forceful on democracy in sub-Saharan Africa, Latin America (especially Haiti), Eastern Europe and the former Soviet Union (Carothers, 1999).

Some commentators castigate the United States for its democracy bandwagon. According to Robinson (1996: 6): 'All over the world, the United States is now promoting its version of "democracy" as a way to relieve pressure from substantive groups for more fundamental political, social, and economic change.' Further, the democratic ideal is hardly evident in the Western countries, with their well-documented flaws, including corruption, favoritism, unequal access to political power, not to mention voter apathy, cynicism and political disengagement (Diamond, 1999; Kaplan, 1997).

Over $700 million is now spent by the United States in promoting democracy, by governmental agencies that are directly involved in the global project. Prominent among these are USAID (US Agency for International Development) and the US Information Agency, while others are government-funded but privately run (Eurasia Foundation, Asia Foundation, and the National Endowment for Democracy). In turn, these private foundations fund other groups like IFES (International Foundation for Electoral Systems), the Carter Center, universities, research institutes, and policy institutes (Carothers, 1999). The breakdown of the funding allocates $147 million for development of legal institutions and law, $203 millions for governance, $230 for civil society, and $60 million for elections and political processes. By geographic region, $87 million was spent in Latin America, $288 million in Eastern Europe and the former Soviet Union, $123 million for sub-Saharan Africa, and $112 million for Asia and the Middle East, with $27 million for unspecified global activities. Altogether, the United States is promoting democracy in over 100 countries (Carothers, 1997). Though in the past, especially during the Cold War, most money was spent on a 'top-down' approach by attempting to boost government institutions, now the strategy is more balanced with a renewed emphasis on 'bottom-up' grassroots democracy and civic organisation.

Despite these impressive numbers, it is unclear how effective democracy promotion is. Whether from the perspective of the recipient or the giver, both of whom have a stake in trying to show the effectiveness of the programs, there is little questioning of the enterprise. A report to the US Congress in 1996 concluded that the US-funded democracy projects in the 1991–96 period in Russia had

> mixed results in meeting their stated developmental objectives [...]. Our analysis indicated that the most important factors determining project impact were Russian economic and political conditions [...]. State (Department) and USAID officials acknowledged that democratic reforms in Russia may take longer to achieve than they initially anticipated. (US General Accounting Office, 1996: 2–3)

Carothers (1999: 59ff) reviews the cavils of the skeptics of the promotion exercise under five headings: (1) rhetoric is more important than substance as the United States supports dictators when it wants; (2) democracy assistance is only a small fraction of US foreign aid; (3) democracy aid is just a pretty way of packaging illegitimate US intervention in the internal affairs of the other countries; (4) democracy cannot be exported and it must be grown from within; and (5) where does the United States get off telling other countries how to run their political systems. Contemporary democratisation requires concessions from those who were formerly excluded from participation, even if the ideals are promoted from outside and are hardly resisted openly from the governing regimes. The main hope is to establish the 'virtuous circle' where stable democratic institutions build civic engagement and trust between individuals and the state.

Conclusion

In the early post-Cold War years, many commentators produced reckless speculations about the benign effects of the spread of democratisation to the majority of the world's countries and, at least, to parts of all the world's regions. But more level-headed analysis took careful note that the number of liberal democracies was not increasing as predicted and had leveled off by the mid-1990s. Like previous waves, the 'Third Wave' could be reversed. What might distinguish this epoch from previous ones is the global hegemonic Zeitgeist of the benefits of liberal democracy and the lack of attraction of any alternative form of government. It is increasingly difficult for a country to remain immune to globalisation influences, including those of a political nature such as the latest wave of democratisation.

Global democratisation after September 11, 2001 has taken on a new energy, at least from the perspective of the United States. After the terrorist attacks, a consistent theme of the Bush Administration is that the installation of democratic regimes in countries from which terror emanates will reduce the chances of a September 11 recurrence. But there is little empirical evidence for this belief. While the causes of terror are complex, it is certainly the case that some groups turn to violence when the outlet for political expression is blocked. Though radical political groups might turn to the ballot box to try to implement their ideologies, the history of such attempts is not one that augurs well for the electoral route to power. In 1992, the Islamicist party, FIS (Islamic Salvation Front) in Algeria was on the verge of an electoral victory when a military coup, supported by France and the United States, pre-empted FIS coming to power via the democratic route. Similar events in Africa, Asia and Latin America have further radicalised ethnic, national and religious movements. In 'illiberal democracies', the unfairness of electoral contests, including restrictions on political mobilisation, party formation, campaigning and access to mass media, have convinced many groups that the odds of having a fair hearing are stacked against them. Consequently, they adopt alternative strategies, including guerrilla tactics. In Central Asia, especially in Uzbekistan, Islamicists have been forced out of the formal democratic political arena by authoritarian tactics from the post-Soviet leaderships. In the frontline of the current 'war on terrorism', these countries are under no serious pressure by the

West to reform. Though the Western pressure for democratisation is packaged well in rhetoric about fairness, civil rights and respect for minority views, the reality is far more complex and numerous instances of continued human rights abuses by Western allies should disabuse anyone of a simple global democratisation trend that is transforming societies everywhere.

In many respects, globalisation in the form of political democratisation is similar to globalisation in the form of economic liberalisation. While the principles behind both trends can be welcomed, the reality is far messier. Economic globalisation is now increasingly challenged by its supposed beneficiaries, as attention turns to the institutions and powerful actors that guide the process to their own benefits and to its unequal impacts. Democratisation is also looked at more skeptically. This chapter has attempted to point out the differences between the various forms of democracy, their distributions, and some of the key reasons why the world political map is changing. The debate about the process is not yet over, and growing skepticism about the nature of democracy as applied in heretofore non-democratic states, despite the gloss and aura of this form of political structure, can be expected to heat up the discussion. The match of ideal and reality will be continually under scrutiny. The questions that remain are: What kind of democratisation and for whom? And who is promoting it and for what purposes?

Notes

The research used in this chapter was supported by the National Science Foundation and was carried out in the context of the graduate training program 'Globalisation and Democratisation' in the Institute of Behavioral Science at the University of Colorado. Special thanks to Mike Ward, now at the University of Washington, and to my other IBS colleagues and students for constructing a stimulating work environment. The maps and graphs were originally prepared by Frank Witmer.

1 The point at which the curve of adopters of the diffusion flattens.
2 See <http://weber.ucsd.edu/~kgledits/Polity.html>.
3 See <http://www.freedomhouse.org/freedom house>.
4 See <http://www.hrw.org/>.
5 The scores for each dimension for each country since 1972 are available from <www.freedomhouse.org/ratings/index.htm>.
6 See <www.usaid.gov/about/>.

References

Agnew, J. A. (1996) 'Mapping politics: How context counts in electoral geography', *Political Geography*, 15 (2): 129–46.

Anderson, L. (ed.) (1999) *Transitions to Democracy*. New York: Cambridge University Press.

Archibughu, D., Held, D. and Kohler, M. (eds) (1998) *Re-Imagining Political Community: Studies in Cosmopolitan Democracy*. Stanford, CA: Stanford University Press.

Beetham, D. (1994) *Defining and Measuring Democracy*. London: Sage.

Bermeo, N. (1999) 'Myths of moderation: Confrontation and conflict during democratic transitions', in L. Anderson (ed.), *Transitions to Democracy*. New York: Cambridge University Press, pp. 120–40.

Bobbio, N. (1989) *Democracy and Dictatorship: The Nature and Limits of State Power*. Minneapolis: University of Minnesota Press.

Boeninger, E. (1997) 'Latin America's multiple challenges', in L. Diamond, M. E. Plattner, Y.-H. Chu and H.-M. Tien (eds), *Consolidating the Third Wave Democracies: Regional Challenges*. Baltimore, MD: Johns Hopkins University Press, pp. 26–63.

Bollen, K. (1993) 'Liberal democracy: Validity and method factors in cross-national measures', *American Journal of Political Science*, 37 (4): 1207–30.

Carothers, T. (1997) 'Democracy without illusions', *Foreign Affairs*, 76 (1): 85–99.

Carothers, T. (1999) *Aiding Democracy Abroad: The Learning Curve*. Washington DC: The Carnegie Endowment for International Peace.

Collier, D. and Levitsky, S. (1997) 'Democracy with adjectives: Conceptual innovation in comparative research', *World Politics*, 49 (3): 430–51.

Dahl, R. A. (1971) *Polyarchy: Participation and Opposition*. New Haven, CT: Yale University Press.

Diamond, L. (1997) 'Introduction: The search for consolidation,' in L. Diamond, M. E. Plattner, Y.-H. Chu and H.-M. Tien (eds), *Consolidating the Third Wave Democracies: Regional Challenges*. Baltimore, MD: Johns Hopkins University Press, pp. xiii–xlvii.

Diamond, Larry J. (1999) *Developing Democracy: Toward Consolidation*. Baltimore, MD: Johns Hopkins University Press.

Dorling, D. (2001) 'Anecdote is the singular of data,' *Environment and Planning A; 33* (8): 1335–69.

Falk, R. A. (2000) *Human Rights: The Pursuit of Justice in a Globalising World*. London: Routledge.

Freedom House (2000) 'Survey methodology', <www.freedomhouse.org/research/freeworld/2000/methodology.htm>

Fukuyama, F. (1992) *The End of History and the Last Man*. New York: Free Press.

Goldman, R. M. and Douglas, W. A. (eds) (1988) *Promoting Democracy: Opportunities and Issues*. New York: Praeger.

Haggard, S. and Kaufman, R. R. (1999) 'The political economy of democratic transitions', in L. Anderson (ed.), *Transitions to Democracy*. New York: Cambridge University Press, pp. 72–96.

Held, D. (1993) 'Democracy: From city-states to cosmopolitan order', in D. Held (ed.), *Prospects for Democracy: North, South, East, West*. Stanford, CA: Stanford University Press, pp. 13–52.

Held, D., Goldblatt, D. and Perraton, J. (1999) *Global Transformations: Politics, Economics and Culture*. Stanford, CA: Stanford University Press.

Huber, E., Rueschemeyer, D. and Stephens, J. (1999) 'The paradoxes of contemporary democracy: Formal, participatory and social dimensions'. in L. Anderson (ed.), *Transitions to Democracy*. New York: Cambridge University Press. pp. 168–92.

Huntington, S. P. (1991) *The Third Wave: Democratisation in the Late Twentieth Century.* Norman, OK: University of Oklahoma Press.

Inglehart, R. and Carbello, M. (1997) 'Does Latin America exist (and is there a Confusian culture)?: A global analysis of cross-cultural differences', *PS: Political Science and Politics*, 30 (1): 34–46.

Joseph, R. (1997) 'Democratisation in Africa since 1989: Comparative and theoretical perspectives', *Comparative Politics*, 29 (3): 363–82.

Kaplan, R. (1997) 'Was democracy a moment?', *The Atlantic Monthly,* December: 55–80.

King, G. (1996) 'Why context should not count', *Political Geography*, 15 (2): 159–64.

Kopstein, J. S. and Reilly, D. A. (2000) 'Geographic diffusion and the transformation of the post-Communist world', *World Politics*, 53 (1): 1–37.

Linz, J. J. and Stepan, A. (1996) *Problems of Democratic Transition and Consolidation: Southern Europe, South America, and Post-Communist Europe.* Baltimore, MD: Johns Hopkins University Press.

Lipset, S. M. (1959) 'Some social requisites of democracy: Economic development and political legitimacy', *American Political Science Review*, 53 (1): 69–105.

Lipset, S. M. (1994) 'The social requisites of democracy revisited', *American Sociological Review*, 59 (1): 1–22.

Lipset, S., M., Seong, K.-R. and Torres, J. C. (1993) 'A comparative analysis of the social requisites of democracy', *International Social Science Journal*, 45 (2): 155–75.

Lynch, M. (2001) 'Globalisation and international democracy', *International Studies Review*, 2 (3): 91–101.

Moon, B. (1996) *The Dilemmas of International Trade.* Boulder, CO: Westview Press.

Most, B. A. and Starr, H. (1989) *Inquiry, Logic and International Politics.* Columbia, SC: University of South Carolina Press.

Norris, P. (1999) 'Conclusions: The growth of critical citizens?', in P. Norris (ed.), *Critical Citizens. Global Support for Democratic Government.* New York: Oxford University Press, pp. 257–72.

O'Loughlin, J. (2000) 'Geography as space and geography as place: The divide between political science and political geography continues', *Geopolitics*, 5 (3): 126–37.

O'Loughlin, J. (2001) 'Geography and democracy: The spatial diffusion of political and civil rights', in G. Dijkink and H. Knippenberg (eds), *The Territorial Factor: Political Geography in a Globalising World.* Amsterdam: Universitet van Amsterdam Press, pp. 77–96.

O'Loughlin, J., Ward, M. D., Lofdahl, C. L., Cohen, J. S., Brown, D. S., Reilly, D., Gleditsch, K. S. and Shin, M. (1998) 'The diffusion of democracy 1946–1994', *Annals of the Association of American Geographers*, 88 (4): 545–74.

Przeworski, A. (1991) *Democracy and the Market: Political and Economic Reforms in Eastern Europe and Latin America.* New York: Cambridge University Press.

Risse, T., Ropp, S. C. and Sikking, K. (eds) (1999) *The Power of Human Rights: International Norms, and Domestic Change.* New York: Cambridge University Press.

Robinson, W. C. (1996) *Promoting Polyarchy: Globalisation, US Intervention and Hegemony.* New York: Cambridge University Press.

Rueschemeyer, D., Stephens, E. H. and Stephens, J. D. (1992) *Capitalist Development and Democracy.* Chicago: University of Chicago Press.

Russet, B. M. (1993) *Grasping at Democratic Peace: Principles for a Cold War World*. Princeton, NJ: Princeton University Press.

Rustow, D. A. (1970) 'Transitions to democracy: Toward a dynamic model', *Comparative Politics*, 2 (3): 337–63.

Schumpeter, J. A. (1947) *Capitalism, Socialism, and Democracy*. New York: Harper Torchbooks.

Schwartzman, K. C. (1998) 'Globalisation and democracy', *Annual Review of Sociology*, 24: 158–81.

Shin, D. C. (1994) 'On the third wave of democratisation: A synthesis and evaluation of recent theory and research', *World Politics*, 47 (1): 135–70.

Siverson, R. and Starr, H. (1991) *The Diffusion of War: A Study of Opportunity and Willingness*. Ann Arbor: University of Michigan Press.

Talbott, S. (1996) 'Democracy and the national interest', *Foreign Affairs*, 75 (6): 47–63.

US General Accounting Office (1996) *Promoting Democracy: Progress Report on US Democratic Development Assistance to Russia*. Washington, DC: US General Accounting Office.

Zakaria, F. (1997) 'The rise of illiberal democracy,' *Foreign Affairs*, 76 (6): 22–43.

3 Electoral Geography in Electoral Studies: Putting Voters in Their Place

Ron Johnston and Charles Pattie

No general textbook on electoral geography has been published for more than two decades,[1] although electoral concerns have received considerable coverage in general political geography texts (e.g. Taylor and Flint, 2000), studies of individual countries (e.g. Shelley et al., 1996), and works on individual elections/ sequences of elections (Archer and Taylor, 1981; Johnston, Pattie and Allsopp, 1988; Johnston et al., 2001a). In *Geography of Elections*, Taylor and Johnston (1979) treated the subject as a linear sequence with three main components, each with its own geography; this was formalized in a series of flow diagrams in a parallel book, *Political, Electoral and Spatial Systems* (Johnston, 1979: 21–23). The *inputs* were the voting patterns – what types of people voted for which parties and/or candidates, where and why; the *throughputs* were the processes which converted votes into a pattern of representation – the geography of the translation from votes to seats, on which Gudgin and Taylor (1979) wrote a classic research monograph; and the *outputs* were the geography of power, addressing questions of who gets what, where and why – of which the geography of the pork-barrel was a key element (Johnston, 1980; Shelley et al., 1996). Similarly, in the first and subsequent editions of *The Dictionary of Human Geography* (1981), Johnston identified five main topics within electoral geography: (1) the spatial organization of elections; (2) spatial variations in voting patterns; (3) environmental and spatial influence on voting patterns; (4) spatial patterns of representation; and (5) spatial variations in power and policy implementation. This framework was little altered when the field was reviewed for the latest edition of the *International Encyclopedia of the Social and Behavioral Sciences* (Johnston, 2001a; see also Johnston, 2001b).

Others have identified several lacunae in the current coverage (see Clark, 1990; Taylor, 1990). The most comprehensive alternative structures were provided by Agnew (1990) and Reynolds (1990). Agnew identified four main objectives in contemporary electoral geography: (1) the geography of electoral behaviour; (2) the effect of the geography of interpersonal information flows on voting behaviour; (3) the geography of electoral systems; and (4) the geography of organization and mobilization by political parties. The first three share common ground with Johnston and Taylor's. The fourth, on which Agnew claimed relatively little had been done, is at best implicit in the point regarding environmental and spatial influences. In his critique, Agnew stressed three main points: (1) geography is usually treated as either epiphenomenal (the outcome of deeper processes operating at larger spatial scales, such as the nation-state) or residual; (2) there is a general, if implicit, belief that empirical findings 'speak for themselves'; and (3) there is an obsession with methodological rather than theoretical discussion. The last point imbues Agnew's entire analysis: electoral geography is largely atheoretical, certainly with regard to geographical/spatial theory, and its main challenge is 'to become more geographical' (Agnew, 1990: 20). His work has built on that argument, with theoretically informed empirical analyses of the geography of voting, notably in Scotland and Italy (Agnew, 1986).

In 1978 Taylor argued that elections are a 'positivist's dream', with a bonanza of data inviting research but unfortunately there had been no 'coherent development of theories and ideas' (1978: 153), resulting from the increasing volume of quantitative studies. More than two decades later, Taylor and Flint (2000: 235) similarly contended that in terms of the volume of work produced, 'electoral geography has been the success story of modern political geography'. But they then ask what this work is 'adding to our knowledge of political geography?'.

> It is not at all clear where electoral geography has been leading. The goal of most studies seems to be nothing more than understanding the particular situation under consideration. The result has been a general failure to link geographies of elections together into a coherent body of knowledge. In short, we have a 'bitty' and uncoordinated pattern of researches, which has produced a large number of isolated findings but few generalizations. (Taylor and Flint, 2000: 236)

Because of this, and also because of the absence of studies of elections in the periphery and semi-periphery of the world economy (on which, see Osei-Kwame and Taylor, 1984; Taylor, 1986), they argue that electoral geography is unable to make credible contributions to 'debates on democratization' at a 'time when the spread of democratic practices across the world provides some hope for humanizing globalization' (see O'Loughlin et al., 1998). Taylor and Flint's agenda is much wider than that essayed here, but their critique is nonetheless important for the basic issues that it raises regarding the wider context within which electoral geography is set.

Although electoral geography's agenda has widened over the last decade, Agnew's third criticism still carries much weight. The field has changed insufficiently for an entirely new framework to be introduced at this stage. But theoretical material does incorporate geography in ways other than those criticized by Agnew; it provides the context for our recasting of the framework here, which focuses on the quantitative empirical exploration of theoretical issues.

A Model for Electoral Geography?

The absence of recent relevant texts is not confined to electoral geography: there is a similar lack in electoral studies more widely, despite the appearance of major books on particular themes (such as Harrop and Miller, 1987; Taagepera and Shugart, 1989; Lijphart, 1994; Cox, 1997; Farrell, 2001). For this overview, we have adapted and extended the approach developed by Miller and Shanks (1996) in their book on *The New American Voter*.

Miller and Shanks' focus is upon accounting for vote choice. They termed their model (hereafter MS model) a *funnel of causation* – although this is never represented diagrammatically – with the voting decision comprising a series of six stages, from the most distant in time (at the funnel's mouth) to that most proximate to the election. As electors move through the stages, their choices become more constrained. In Figure 3.1, we have added an additional stage after the voter choice that involves the translation of vote into the election outcome.

The first stage of the MS model incorporates sociological models of voting, which portray electoral choice as a function of voters' individual characteristics – indexed by such variables as parents' class, parents' political preferences, and respondents' class, education, housing tenure, and union membership. These factors underpin the development of personal ideologies, the principles on which responses to electoral stimuli are based (Scarbrough, 1984). They act as the foundations on which attitudes (more short-term and empirically-based than values) are formed and may lead voters to develop long-term identifications with particular political parties.

In classic *sociological models*, the variables identified in stages 1–2 of the MS model lead directly to voter choice. Once their political identities are established, many habitually vote in the same way at a sequence of elections – what classic studies of the American voter termed the 'normal vote'. Some do not, however, either because they have not developed a stable identification with any one party or because contemporary events cause them to deviate from their 'normal' voting pattern (as enunciated in Key's (1955) classic theory of critical elections). The sociological model has increasingly been found wanting, however: some dismiss it entirely; others see its main variables lacking the quantitative predictive ability of previous electoral epochs. (The emphasis on class reflects work in the English-speaking countries. Elsewhere in Europe, Dogan (2001: 112) claims that 'it was religion and everything which comes with it ... that has predominated in electoral behaviour, and not social class'.) Voters are increasingly open to alternative claims for their support, and iden-tification with individual parties has become weaker, a process called partisan dealignment. Their decisions are largely influenced by evaluations of current conditions (dominantly but not solely economic) and the perceived perfor-mance of parties and candidates in government, by their expectations regard-ing the future, and by their assessments of the potential governance offered by rivals. Their behaviour is the subject of *responsive voter* models, which occupy stages 3–6 of the MS model.

Responding to the various salient issues at the election, in the context of their socialized political values and attitudes, electors make their choices in a threefold sequence (see Figure 3.1): whether to vote or abstain (an increasingly important decision, given recent turnout decline in many countries); who to vote for; and if

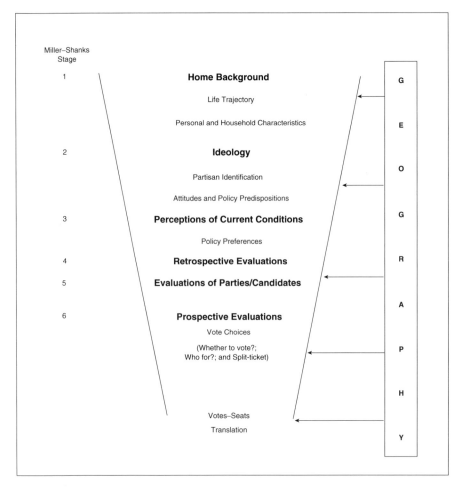

Figure 3.1   Geography and the funnel of causation

relevant, as is increasingly the case in many electoral systems, whether to vote in different ways in different components of the election (the split-ticket decision). Those choices reflect a complex interplay of two major sets of processes: *political socialization*, the processes by which the principles underlying behaviour are learned; and *political mobilization*, the processes of vote seeking by interested parties.

Extending the Model: Bringing Geography In

Although we use the MS model as a framework, three basic criticisms stimulate suggestions for its expansion. First, the model presents a linear sequence, but not all voters will go through this at every election; some stages may be jumped and some potential influences ignored (even unknown). Furthermore, there will be *feedback loops*. Political socialization is continual, even if it becomes slower and less effective as people age: older people are more set in their ways than younger ones, and also less likely to change their basic socio-economic characteristics, such as housing tenure and educational achievements. Additionally, political mobilization efforts immediately before an election, when voters evaluate the current situation, may lead to reassessments of political attitudes, even ideology, and hence of partisan identification.

Secondly, the model largely ignores some key players, notably the *political parties* and others seeking power. It assumes a responsible electorate making informed choices. But where do electors get information for such choices, especially in the responsive voter stages? Many rely for their sources on those immediately most affected by the choices – the political parties and their candidates, plus associated interest groups. Much rational choice-inspired research (derived from Downs, 1957) shows that acquiring information takes time and is costly, with the benefits often incommensurate with those costs. So many voters depend on freely provided, readily accessible information from those with money to spend on campaigning and canvassing, and by others with vested interests, such as media organizations.

Finally there is *geography*. Figure 3.1 suggests that geography influences all stages of the sequence, plus the later processes of translating votes into seats. However, this does not imply, in Agnew's terms, that geography is either an epiphenomenal or a residual influence – either a superficial representation of a deeper set of influences (i.e. geography is merely a surrogate for them) or a trivial influence, to be taken into account when all others have been considered. Critics of electoral geography (MacAllister and Studlar, 1992) have attacked it on both grounds. The response underpinning this chapter is that geography is a real effect at all stages of the MS model as set out in Figure 3.1, and also of the feedback loops it omits.

Such criticisms are based on a myopic view of social processes, brought into stark contrast by Thrift's (1983) division of theories of individual behaviour into the compositional and the contextual. *Compositional theories* characteristically locate all fundamental causes of behaviour in the individuals themselves and the socio-economic categories to which they are allocated: they are atomistic accounts, placing individuals in social categories without geographical/spatial referents, save perhaps the nation-state. *Contextual theories*, on the other hand, place individuals in their social milieu, many of which are locales, spatially defined settings for interaction where learning occurs. These two are ideal types: it is extremely difficult to understand individuals without also appreciating the places in which they learn what it means to belong to a particular social category; and it is equally difficult to put people in places without any other social categories to characterize them. Compositional and contextual theories are necessarily interdependent: we are what we are, not only because of who we are but also because of where we have been, and where we are now. Learning is very much (though far from completely) a place-bound process involving interpersonal interaction.

This case could be interpreted as trivial, in line with the residual and epiphenomenal arguments: geography is just the decision-making arena. But this portrays it as a setting with no causative functions. Structuration theory (Giddens, 1984) argues that structure and agency are constantly interacting and producing change in each other. And geographical settings are structural components in such a dialectic. Places are not inert settings, they have cultures: shaped by people, significant for the shaping of people, and then re-shaped by people in a continuing dialectic. Geographies are continually being created and re-created as part of the ongoing processes of socialization and mobilization, and so are the inhabitants of these geographies.

Within electoral studies, therefore, electoral geographers ensure that a geographical – or spatial, or contextual – perspective is central to the evolving understanding of how voters make choices and electoral systems operate. Place matters in those processes. Various approaches illustrate that claim. This chapter focuses on the quantitative, illustrating both the achievements and the methodological and substantive issues still high on the research agenda.

Place Matters: The Geography of Political Socialization

The compositional model has dominated sociological approaches to electoral studies, strongly influenced by Lipset and Rokkan's (1967) classic essay on electoral cleavages in Western Europe (see also Bartolini, 2000; Karvonen and Kuhnle, 2001). They argued that most countries' basic electoral alignments were in place by 1930, as parties mobilized support on either side of one or more social cleavages – potential divisions within society resulting from the national and industrial revolutions. Some countries have complex party systems and voting behaviour reflecting the continued presence of several such cleavages and the ease of accommodating them within the electoral system; others are dominated by a single – usually class – cleavage (with Great Britain a paradigm exemplar).

These approaches assume that voting behaviour can be 'read off' once individuals' positions relative to the known cleavages are established. This 'national compositional voting model' position has geography as epiphenomenal: if class is the key cleavage, knowledge of peoples' class positions provides immediate access to their political attitudes and voting choices – irrespective of where they live – and any geography of party support will simply represent the geography of class membership. This oversimplistic position was countered by sensitive studies such as Butler and Stokes' (1969; 1974) path-breaking British work, which demonstrated within-class variations in party choice readily associated with different social milieu. A member of the working class in a city dominated by large enterprises, where trades unions were very active, is probably politically socialized in a very different environment from somebody in a similar class position living in a small town or in the countryside; the latter would assimilate very different conceptions of what it meant to be in the working class, of attitudes to power and authority, and of which political party they should support. This was the case, for example, in the US South for much of the twentieth century, where whites were socialized into a particular set of attitudes towards blacks, and to an appreciation (usually uncontested) that white supremacy could be sustained by voting Democrat (rather than for the Republican party, associated with the Northern cause in the Civil War: see Key, 1949; Wright, 1977).

Despite such findings, however, such geographically-variable, within-class differentials are largely treated as residuals, giving geography only a marginal influence. But is that so? Answering this question posed two related issues. Empirical demonstrations of geographical variations in voting behaviour are necessary but not sufficient: conversion to the importance of a geographical perspective requires it to be theoretically grounded.

Theoretical underpinnings were central to the early burgeoning of electoral geography, led by Cox (1969a) and Reynolds' (1969) generalization of Key's (1949) friends-and-neighbours model. Political learning involves interpersonal interaction – at home, school and college, at the workplace, and in the local neighbourhood and formal organizations, for example. Information flows through such social networks are the core of structuration processes. Places acquire their own political cultures based on these, which may vary from those of otherwise similar places. As a consequence, they may display greater political and electoral than social polarization because the majority view prevails in such interaction, with more people converted to it than to any minority position (assuming equal, or random, willingness to be persuaded). Thus in Britain, with a predominantly pro-Labour working class and predominantly pro-Conservative middle class, heavily working-class areas have even more Labour supporters/voters than their class composition suggests, with the opposite situation in strongly middle class areas. This comes about via 'conversion through conversation' (or 'those who talk together, vote together': Miller, 1977: 67), whereby, for example, because they are predominantly exposed to pro-Labour information there, the more working-class an area then not only the more members of the local working class who vote Labour but also the more members of the minority middle class who vote for that party too.

Empirical Evaluation: Neighbourhood Effects

Empirical evaluation of this hypothesis has been difficult to assemble. Ideally, survey data about individuals should be combined with information characterizing their social milieux. But after Cox's (1969b) original work, little geographical work employed survey data until the 1990s, and instead relied greatly on aggregate, or ecological, data. The latter present important difficulties in the interpretation of data: for example, by showing that areas with large relative numbers of miners living in them also have large percentages voting for a left-wing party does not necessarily mean that miners vote for left-wing parties. This is an ecological fallacy (Alker, 1969), of inferring an individual relationship from an aggregate one, whose resolution has faced social scientists with a major test, to which King (1997) has recently claimed an answer, a method which estimates the individual relationship from the aggregate data (see also Sui et al., 2000). King's method has been used in studies of voting for the Nazi party in the 1930s Germany (O'Loughlin, 2000).

Among geographers, a pioneering proposal for resolving this fallacy involved combining survey and aggregate data, with which it was straightforward to demonstrate the invalidity of the 'national pattern' assumption. UK survey data established the pattern of class voting across the country as a whole, for example; census data showed the class composition of each Parliamentary constituency for which election returns were available. If the 'national pattern' assumption held, then the election outcome in each constituency should be predicted from knowledge of its class composition. But there are very substantial disparities between the predicted and actual results (Johnston, Pattie and Allsopp, 1988), clearly indicating that similar people (i.e. in the same class category) did not vote in the same way wherever they lived.

But did this finding confirm the geographical polarization thesis? Answering this required data on how similar people voted in different places. Estimates for these, consistent with the known data on class composition and election outcomes, plus the national pattern obtained via a survey, can be obtained using *entropy-maximizing* (Johnston and Pattie, 2000). For Great Britain in 1979–1987, these estimates indicated very significant within-class variations, and clearly showed, for example, that the more working-class the constituency the more members of each class who voted Conservative (Johnston, 1985; Johnston, Pattie and Allsopp, 1988).

This estimating procedure was also deployed to demonstrate that the degree of polarization varied regionally, especially over the 1979–1987 period (Johnston, Pattie and Russell, 1993). One major consequence of Thatcher's economic policies was rapid and massive decline in the UK's manufacturing industries, concentrated in the urban areas and in the North of the country, countered by significant growth in the service sector, concentrated in the South. The already-existing North–South divide opened up very substantially, with the class divide closing in the South (where the prospering working class increasingly voted Conservative) but widening in the North. Here we see the feedback loop between the sociological and responsive voter models: the pattern of class voting changed as a consequence of the country's changing economic geography.

These findings are consistent with hypotheses regarding geographical polarization, but say nothing about the processes involved (Dunleavy, 1979). They also have a problem regarding their *spatial scale*. Parliamentary constituencies (with populations approaching 100,000) are much larger in area than the assumed neighbourhoods through whose social networks electorally-relevant information flows. Recent small-scale studies have placed individuals' voting decisions firmly in their local contexts, however, using small-area census data to characterize the milieux for each of 2,730 respondents to the 1997 British Election Study survey. These bespoke neighbourhoods for each voter identified the main characteristics of the people living closest to their home addresses (the nearest 1,000, say, or the nearest 2,500) and enabled identification of clear neighbourhood effects: the more middle-class the neighbourhood, for example, the greater the level of Conservative voting in each of the three main social classes (MacAllister et al., 2001), whatever other individual characteristics were held constant, and the more unemployed living locally, the more people voted against the governing party seen as responsible for that unemployment (Johnston et al., 2000, 2001a, 2001b). Geography is not epiphenomenal: it provides the arena (at a variety of scales) within which electoral decisions are made.

Strong circumstantial evidence from these studies using bespoke neighbourhoods confirms the neighbourhood effect hypothesis, but they lack information on process – on whether '*conversion by conversation*' occurs. Few studies have approached this problem directly (see Eagles, 1995a, 1995b), and only one has collected a large data set that addresses the process issue (Huckfeldt and Sprague, 1995). This study of a single town in the USA found that people who talk to their neighbours are more likely to vote in the same way as they do, especially if they are neighbours whom they met through the formal structures of local church organizations (Huckfeldt, Plutzer and Sprague, 1993). Similarly, British data show that 'people who talk together do vote together' (Pattie and Johnston, 2000), though without evidence that the conversations were locally-focused, but clearly indicating, as Cox (1969b) had suggested, that those with weak partisan leanings were more likely to change their votes than those strongly associated with one party (Pattie and Johnston and Pattie, 2002a). We are getting closer to a complete test, but are not there yet.

Economic voting and local context

Whereas the 'conversion through conversation' hypothesis is a relatively convincing argument for neighbourhood effects in the context of sociological models of voting behaviour, it can also apply to the responsive voter hypothesis. Voters responding to economic and other conditions may rely on a variety of cues and stimuli, one set of which come from their friends and relations while others come from their observations of local conditions (Books and Prysby, 1993).

Economic voting models relate people's retrospective and prospective evaluations of economic conditions, and government responsibility for them, to voting choices: they provide compelling evidence that governments which deliver prosperity tend to be rewarded by being returned to power whereas those blamed for economic recession/depression are punished by being voted out of office (Lewis-Beck and Stegmaier, 2000). Such studies usually look at evaluations of changes in the national economic condition and in the individual voter's situation. And yet, whatever the gross pattern over the country, different areas may have very different experiences. British voters independently evaluate the economic situation not only nationally and individually but also locally, and make their voting decisions accordingly (Pattie, Dorling and Johnston, 1997; Johnston and Pattie, 2001). Furthermore, those evaluations are especially influential if a voter's local area has a high level of unemployment (the subjective and objective interacted to produce high levels of anti-government voting in 1997: Johnston et al., 2000b; Tunstall et al., 2000). This suggests that there are altruistic voters, who vote against their apparent individual self-interest if that is contrary to the situation that many of their neighbours face (Johnston, Dorling et al., 2000).

Where Next Methodologically?

These recent approaches to understanding voting maps – of who votes what, where, at a variety of spatial scales – have used novel methodological procedures to identify the basic patterns. Most have involved moving away from a heavy reliance on aggregate spatial data, with the attendant problems associated with the ecological fallacy, and have involved geographers analyzing individual-level survey data. Because of their concern with spatial variations, however, and with geography as a context for electoral decision-making, they have explored ways of integrating individual and aggregate data in order to test their hypotheses. This has involved adopting methods for analyzing nominal data, and of the interactions between individual and local area characteristics (Russell, 1997), as in studies of how people with different educational levels respond to various electoral stimuli (Pattie and Johnston, 2001).

The theoretical appreciation that place may matter in different ways in different locales, reflecting local cultures, histories, and conditions, calls for analytical procedures which can decompose spatial variations. An approach that can achieve this is multi-level modelling (Jones, 1991), which explores the relationship between two variables across different places (or types of place). An early application found different relationships between the percentage of miners in a constituency and Labour voting in 1987 in separate UK regions, reflecting different responses to the 1984–1985 miners' strike (Jones, Johnston and Pattie, 1992): later work (Jones, Gould and Watt, 1998) showed that it is not only what you are (indexed by class and age, for example) and what you have (housing tenure and employment status) but also where you live (what region, what type of area) that influences how you vote.

Bringing the Parties in: Place and Political Mobilization

Parties and their candidates are major providers of politically and electorally relevant information. Although their campaigns are especially intensive immediately prior to an election, many now operate almost continuous campaigns which are buttressed in the final weeks by parties' canvassing efforts designed to identify their own supporters, sustain that support where it is wavering, and then ensure that they turn out on election day.

Much campaigning employs mass media, especially where parties can buy airtime, underpinned by locally organized efforts to identify and win over support. Much canvassing in the UK, for example, is done by party activists visiting voters on their doorsteps, although there is increasing use of the telephone (Denver and Hands, 1997). Despite some claims to the contrary in the UK, evidence building on American findings (e.g. Jacobson, 1978) has strongly indicated that local campaigns are effective. This is because: (1) the more activists a party has in a constituency prepared to campaign and canvass for votes, the better its performance (Whiteley and Seyd, 1994); (2) the more intensive a party's local campaign, the better its performance (Denver and Hands, 1997); and (3) the more that a local party spends on a constituency campaign (mainly on posters and leaflets), the better its performance there (Pattie, Johnston and Fieldhouse, 1995). These three approaches measure different aspects of local campaign intensity (Pattie et al., 1994), but each has its own separate impacts (on spending, see Johnston and Pattie, 1996). People canvassed by a party are more likely to switch their vote to it than those who are not (Fieldhouse et al., 1996), for example, especially if the canvassing is personal, on the doorstep (Gerber and Green, 2000; Pattie and Johnston, 2003).

Such campaigns are certainly effective where parties are promoting tactical voting. In Britain's three-party system (four-party in Scotland and Wales), over recent decades the main goal for some voters may have been either to remove the incumbent government or to prevent victory by another party, and they may have been convinced to vote for the party best-placed in their constituency to achieve that goal, even if it was not their first choice on other grounds. This is known as tactical voting in the UK, and as strategic voting in the USA (Cox, 1997). In 1997, for example, defeat of the Conservatives was a primary objective for many voters. Pre-election polls suggested that Labour was going to win overall, and some Labour supporters were persuaded to vote for the Liberal Democrat candidate (or nationalist party in Scotland or Wales) if that person was best placed in their constituency to unseat a Conservative incumbent. Many Liberal Democrats campaigned on that argument in constituencies where it was to their advantage. Estimates of tactical voting (Johnston and Pattie, 1991; Alvarez and Nagler, 2000) by constituency are consistent with these expectations: more took place in marginal than safe Conservative-held seats, for example, especially where the parties involved campaigned intensively: tactical voting was greatest where it was most likely to be efficacious and where the parties campaigned hardest for it (Johnston, Pattie et al., 1997; Johnston, Pattie, Dorling and Rossiter, 2001a).

Most work on local campaigns has focused on the first two of the three decisions identified in Figure 3.1: whether to vote, and, if so, for which party/candidate. This assumes a single election, and a single vote decision. In the USA, however, voters in Presidential contests have to make decisions in parallel elections to the House of Representatives, possibly to the Senate, and also for a wide range of State and local positions. They must decide whether to vote for the same party in all contests, or whether to vote a split-ticket, supporting one party's candidate for one office but another's for other positions. Recently, a number of countries (Italy, Japan and New Zealand, for example) have adopted the Additional Member (AM) or Mixed-Member Proportional (MMP) system, used in Germany since 1951, with the system also adopted for the Parliament in Scotland and the Assembly in Wales. Some members are elected from single-member constituencies in first-past-the-post contests and the remainder from national or regional party lists, selected so as to ensure overall proportional representation (Shugart and Wattenberg, 2001); voters in such situations can also vote split-tickets, supporting different parties at the constituency and list contests.

Is there a geography to split-ticket voting? Hypotheses from rational voter theory suggest it is greatest where the parties campaign for it most intensively, by providing more information encouraging separate decisions in the two rounds of voting. Analyses of three elections in New Zealand, Scotland and Wales showed that the more money spent by a party in its constituency campaign, the more split-ticket votes that it attracted (Johnston and Pattie, 2002; see also several of the chapters in Shugart and Wattenberg, 2001). Parties operating in places influenced the election outcome, a conclusion reached by Burden and Kimball's (1998) analyses of split-ticket voting in the USA.

Translating Votes into Seats

A perennially intriguing feature of electoral systems is how they translate votes into seats. Legislative elections using the first-past-the-post system are not only most likely to produce disproportionate results but also the most likely to result in: (1) 'manufactured majorities' (no party wins a majority of the votes but one is very likely to be allocated a majority of the seats); (2) 'counter-intuitive' outcomes (the party with the largest number of votes does not receive the largest number of seats); and (3) the most 'responsive' reactions to changes in seat relative to vote distribution, with small changes in vote share generating much larger changes in seat share (Rae, 1967; Taagepera and Shugart, 1989; Lijphart, 1994). Gudgin and Taylor (1979) showed that, even in the absence of any partisan manipulation of constituency boundaries leading to either or both of malapportionment and gerrymandering, the procedure of superimposing a mesh of constituencies over a punctiform geography of voter support for different parties was virtually certain to result in a disproportional outcome. Indeed, in the UK, where independent Boundary Commissions undertake regular, non-partisan reviews of constituency boundaries, there is no fair way to define districts and 'neutral Boundary Commissioners are likely, by the law of averages, to act in the same way as would a gerrymanderer acting for a local majority party. In effect, the British Boundary Commissioners gerrymander for Labour in the inner cities and for the Conservatives in most other areas' (Gudgin and Taylor, 1979: 203: see also Rossiter et al., 1999). Geography is central to the votes-to-seats translation process in all circumstances, and not just in situations where malapportionment and gerrymandering are deliberate strategies: malapportionment was ruled unconstitutional in the 1960s, so all constituencies within a territory must be of the same size, but certain types of gerrymandering, such as attempts to create minority–majority districts (Lennertz, 2000; Forest, 2001), have been found acceptable by the Supreme Court in certain circumstances.

Does such non-partisan gerrymandering treat all parties equally, and if not, how extensively does it influence election outcomes? Although plenty of evidence has been adduced for the existence of both major abuses in electoral cartography, their impact on election results has rarely been analyzed in rigorous, quantitative ways. Disproportionality is the gerrymanderer's goal, seats:votes ratios unfavourable to its opponents. But disproportional election results are the norm with most constituency-based electoral systems (Tufte, 1973; Farrell, 2001). King (1989; King and Browning, 1987; Gelman and King, 1994) argues that the impact of a set of constituencies on election results should be evaluated on two criteria: (1) its *responsiveness* – what is the relationship between the rates of change in a party's percentage of the votes cast and of its percentage of the seats allocated?; and (2) its *bias* – are the parties treated equally, or does the degree of disproportionality (and the rate of responsiveness) vary between them? The latter question has been evaluated by estimating the outcome in terms of seats won if the two main parties got the same share of the votes cast. If one would obtain more seats than the other, then the map is biased towards it.

Any observed bias has to be decomposed, to investigate its origins. Brookes' (1960) methodology measures bias as the number of seats difference between two parties when they have the same vote share, plus a way of 'unpicking the map'. This has been adapted for studies of UK elections between 1950 and 1997, involving five separate redistrictings by the Boundary Commissions and 14 general elections (Johnston, Pattie, Dorling and Rossiter,

2001a; Johnston, Pattie, Dorling and Rossiter, 2001b, extends the study to the 2001 election; see also Blau, 2001). Six bias components are identified. Those associated with *malapportionment* arise: (1) because of differences between England, Scotland, Wales and Northern Ireland in their national electoral quotas (their average constituency electorates); and (2) because of differences within each of those countries in *constituency size* (which arise partly because exact equality is never achieved when constituencies are redefined and partly because of *creeping malapportionment* – changes in constituency size because of population changes). Those associated with *gerrymandering* arise because (3) one party's vote distribution across the constituencies is more efficient than the other's because it wastes fewer in seats that it loses and/or has fewer surplus votes in those that it wins. Those associated with *reactive malapportionment*, in which the number of votes necessary for victory in a constituency is reduced, arise because of: (4) differences between constituencies in the number of abstentions; (5) differences between constituencies in the number of votes for third parties; and (6) the relative strength of the two main parties in constituencies where there are third party victories.

The volume of bias in UK general election results increased substantially over the 15 elections, favouring the Conservatives in the 1950s but Labour from the 1990s on – in part because of successful geographical strategies by Labour, both in its attempts to influence the process of redistribution at the public consultation stage (explicit gerrymandering) but also its targeted constituency campaigns, including those for tactical voting discussed above. In 1997, Labour would have won 82 more seats than the Conservatives with equal vote shares, and in 2001 the bias would have been 141 seats, with the increase between the two elections largely resulting from the lower turnout (which fell most in Labour-held seats) and a greater pro-Labour gerrymander. (Note that we are not referring to gerrymandering in the 'classic' sense of redrawing constituency boundaries for partisan gain – there was no such redistribution between 1997 and 2001 – but rather greater efficiency in the geography of Labour's votes within the given set of constituencies.)

According to this method of establishing bias, at the 2000 US Presidential election, George W. Bush would have had a 50-vote advantage in the Electoral College if he and Al Gore had won equal vote shares at 48.171%. Of that 50-vote advantage, the gerrymander component would have been responsible for some 38, with the malapportionment component adding another 15 in Bush's favour. Only the third party reactive malapportionment component favoured Gore (by just two Electoral College votes), and that was a double-edged sword because if Green Party candidate Ralph Nader had performed less well in the States where he got the best results, and Gore had picked up more votes than Bush as a consequence (especially in Florida and New Hampshire), then Gore would probably have won in the Electoral College as well as the popular vote (Johnston, Pattie and Rossiter, 2001). Bush won because the geography of his voting strength, relative to the State boundaries, produced a biased outcome in his favour: the geography of his votes was more efficient than was Gore's.

Conclusion

Electoral geography has played a peripheral role in both electoral studies and political geography over the last four decades. With regard to its interrelationships with electoral studies, although work by some electoral geographers has been published in journals devoted to political science in general, as well as to specialist electoral journals, nevertheless the benefit of (or even the necessity of) a geographical approach has not been widely appreciated. The model employed here, which builds on one of the main frameworks employed in electoral study, has illustrated the importance of appreciating geography in developing the understanding of the vote-decision process and its aftermath – the translation of votes into seats.

There is evidence that this argument is now taking hold, not least in a series of papers in a special issue of *Electoral Studies* on the future of election studies. In their editorial introduction, Wleizen and Franklin (2002) note the growing interest in contextual effects, leading to a group of three papers on 'addressing space'. Those three (Marsh, 2002; Johnson, et al., 2002; Stoker and Bowers, 2002) concentrate on methodological issues, but also point to some of the conceptual problems associated with contextual effects, such as the extent to which individuals can select their residential contexts. Interestingly, the editors' conclusion (Franklin and Wleizen, 2002) notes that many of the various influences on voting behaviour considered are geographically based, suggesting that exploration of where (and how) people make their voting decisions could become central to future electoral studies. Bringing space in means bringing geography in, though references to the work of geographers are very sparse in the papers included in their collection.

Within electoral geography, the major focus of criticism, as outlined at the beginning of this chapter, has been on the relative balance of theory and empirical work, and the lack of attention to some of the major actors in the electoral process. Most of those criticisms were mooted a decade ago, and since then much work has addressed them (implicitly if not explicitly). The stronger links being forged between electoral geography and electoral studies include essays into theories of space and place (Johnston, 2001c), as illustrated here, and there has been a growing emphasis on how political parties and other actors manipulate spaces and places in a variety of ways (Johnston, 2002).

The most recent criticisms concerning relationships between electoral and political geography (Taylor and Flint, 2000) are intriguingly set in a very positivist framework, arguing that electoral geography has no clear intellectual strategy beyond the conduct of empirical inquiry, and therefore is not contributing a 'coherent body of knowledge' which can inform wider debates regarding democracy and its ever-wider implementation. The immediate response to this is that in electoral studies, as much as in any other area of empirical work in the social sciences, it is necessary to separate the general from the particular, and to realize that most of the empirical findings produced by electoral geographers reflect contingent conditions. This does not preclude generalizations, especially at the aggregate or ecological level, but makes it clear (as realist theories of science have stressed for some time) that general forces involved in voting decisions by individuals, and in attempts to influence those decisions by political actors, are strongly mediated by contingent circumstances. Thus, electoral geographers have shown that political parties manipulate space in a variety of ways to influence electoral outcomes, and that electors respond to these often countervailing forces within their particular contexts, which for many are place- as well as temporally-bounded. Geography matters in the conduct and outcome of elections, but how it matters at any one election (or even at any one place within an election) reflects the circumstances of there and then.

Note

1   The latest in English was Taylor and Johnston (1979).

References

Agnew, J. A. (1986) *Place and Politics*. Boston: Allen & Unwin.

Agnew, J. A. (1990) 'From political methodology to geographical social theory? A critical review of electoral geography, 1960–1987', in R. J. Johnston, F. M. Shelley and P. J. Taylor (eds), *Developments in Electoral Geography*. London: Croom Helm, pp. 15–21.

Alker, H. R. (1969) 'A typology of ecological fallacies', in M. Dogan and S. E. Rokkan (eds), *Quantitative Ecological Analyses in the Social Sciences*. Cambridge, MA: MIT Press, pp. 69–86.

Alvarez, R. M. and Nagler, J. (2000) 'A new approach for modelling strategic voting in multiparty elections', *British Journal of Political Science*, 30, 57–76.

Archer, J. C. and Taylor, P. J. (1981) *Section and Party*. Chichester: John Wiley.

Bartolini, S. (2000) *The Political Mobilization of the European Left, 1860–1980: The Class Cleavage*. Cambridge: Cambridge University Press.

Blau, A. (2001) 'Calculating partisan bias in British general elections without using swing', in P. Cowley et al. (eds), *British Elections and Parties Review* (Volume 11). London: Frank Cass.

Books, C. L. and Prysby, J. W. (1993) *Political Behavior and the Local Context*. New York: Praeger.

Brookes, R. H. (1960) 'The analysis of distorted representation in two-party, single-member elections', *Political Science*, 12, 158–167.

Burden, B. and Kimball, D. (1998) 'A new approach to the study of ticket splitting', *American Political Science Review*, 92, 533–544.

Butler, D. E. and Stokes, D. (1969) *Political Change in Britain*. London: Macmillan.

Butler, D. E. and Stokes, D. (1974) *Electoral Change in Britain*. (Second edition). London: Macmillan.

Clark, G. L. (1990) 'Regulating union representation elections: towards a third type of electoral geography', in R. J. Johnston, F. M. Shelley and P. J. Taylor (eds), *Developments in Electoral Geography*. London: Croom Helm, pp. 242–256.

Cox, G. W. (1997) *Making Votes Count*. Cambridge: Cambridge University Press.

Cox, K. R. (1969a) 'The voting decision in a spatial context', *Progress in Geography 1*. London: Arnold, pp. 81–117.

Cox, K. R. (1969b) 'The spatial structuring of information flow and partisan attitudes', in M. Dogan and S. E. Rokkan (eds), *Quantitative Ecological Analysis in the Social Sciences*. Cambridge, MA: MIT Press, pp. 343–370.

Denver, D. T. and Hands, G. (1997) *Modern Constituency Electioneering: The 1992 General Election*. London: Frank Cass.

Dogan, M. (2001) 'Class, religion, party: triple decline of electoral cleavages in Western Europe', in L. Karvonen and S. Kuhnle (eds), *Party Systems and Voter Alignments Revisited*. London: Routledge, pp. 93–114.

Downs, A. (1957) *An Economic Theory of Democracy*. New York: Harper.

Dunleavy, P. (1979) 'The urban basis of political alignment', *British Journal of Political Science*, 9, 409–443.

Eagles, M. (ed.) (1995a) *Spatial and Contextual Models in Political Research*. London: Taylor and Francis.

Eagles, M. (ed.) (1995b) *Spatial and Contextual Models of Political Behaviour*. *Political Geography*, 14 (6/7).

Farrell, D. M. (2001) *Electoral Systems*. London: Palgrave.

Fieldhouse, E. A., Pattie, C. J. and Johnston, R. J. (1996) 'Tactical voting and party constituency campaigning at the 1992 General Election in England', *British Journal of Political Science*, 26, 403–418.

Forest, B. (2001) 'Mapping democracy: racial identity and the quandary of political representation', *Annals of the Association of American Geographers*, 91, 143–166.

Franklin, M. and Wleizen, C. (2002) 'Reinventing electoral studies', *Electoral Studies*, 21, 331–338.

Gelman, A. and King, G. (1994) 'A unified method of evaluating electoral systems and redistricting plans', *American Journal of Political Science*, 38, 514–554.

Gerber, A. S. and Green, D. P. (2000) 'The effects of canvassing, telephone calls and direct mail on voter turnout: a field experiment', *American Political Science Review*, 94, 653–663.

Giddens, A. (1984) *The Constitution of Society*. Cambridge: Polity Press.

Gudgin, G. and Taylor, P. J. (1979) *Seats, Votes and the Spatial Organisation of Elections*. London: Pion.

Harrop, M. L. and Miller, W. L. (1987) *Elections and Voters*. London: Macmillan.

Huckfeldt, R., Plutzer, E. and Sprague, J. (1993) 'Alternative contexts of political behavior: churches, neighborhoods and individuals', *Journal of Politics*, 55, 365–381.

Huckfeldt, R. and Sprague, J. (1995) *Citizens, Politics and Social Communication: Information and Influence in an Election Campaign*. Cambridge: Cambridge University Press.

Jacobson, G. C. (1978) 'The effects of campaign spending on Congressional elections', *American Political Science Review*, 72, 469–491.

Johnson, M., Shively, W. P. and Stein, R. M. (2002) 'Contextual data and the study of elections and voting behavior: connecting individuals to environments', *Electoral Studies*, 21, 219–234.

Johnston, R. J. (1979) *Political, Electoral and Spatial Systems*. Oxford: Oxford University Press.

Johnston, R. J. (1980) *The Geography of Federal Spending in the United States of America*. Chichester: John Wiley.

Johnston, R. J. (1981) 'Electoral geography', in R. J. Johnston et al. (eds), *Dictionary of Human Geography*. Oxford: Blackwell, pp. 99–100.

Johnston, R. J. (1985) *The Geography of English Politics*. London: Croom Helm.

Johnston, R. J. (2001a) 'Electoral geography', in N. J. Smelser et al. (eds), *International Encyclopedia of the Social Sciences*. Rotterdam: Elsevier, pp. 4374–4378.

Johnston, R. J. (2001b) 'Electoral geography', in J. Michie (ed.), *Reader's Guide to the Social Sciences* (Volume 1). London: Fitzroy Dearborn, pp. 454–456.

Johnston, R. J. (2001c) 'Out of the 'moribund backwater': territory and territoriality in political geography', *Political Geography*, 20, 677–694.

Johnston, R. J. (2002) 'Manipulating maps and winning elections: measuring the impact of malapportionment and gerrymandering', *Political Geography*, 21, 1–32; 55–66.

Johnston, R. J., Dorling, D. F. L., Tunstall, H., Rossiter, D. J., MacAllister, I. and Pattie, C. J. (2000) 'Locating the altruistic voter: context, egocentric voting and support for the Conservative party at the 1997 general election in England and Wales', *Environment and Planning A*, 32, 673–694.

Johnston, R. J. and Pattie, C. J. (1991) 'Tactical voting in Great Britain in 1983 and 1987: an alternative approach', *British Journal of Political Science*, 21, 95–108.

Johnston, R. J. and Pattie, C. J. (1996) 'The value of making an extra effort: campaign spending and electoral outcomes in recent British general elections – a decomposition approach', *Environment and Planning*, 28, 2081–2090.

Johnston, R. J. and Pattie, C. J. (2000) 'Ecological inference and entropy-maximizing: an alternative estimation procedure for split-ticket voting', *Political Analysis*, 8, 333–345.

Johnston, R. J. and Pattie, C. J. (2001) 'It's the economy, stupid' – But which economy? Geographical scales, retrospective economic evaluations and voting at the 1997 British general election', *Regional Studies*, 35, 309–319.

Johnston, R. J. and Pattie, C. J. (2002) 'Campaigning and split-ticket voting in new electoral systems: the first MMP elections in New Zealand, Scotland and Wales', *Electoral Studies*, 21, 583–600.

Johnston, R. J., Pattie, C. J. and Allsopp, J. G. (1988) *A Nation Dividing? Britain's Changing Electoral Map, 1979–1987*. London: Longman.

Johnston, R. J., Pattie, C. J., Dorling, D. F. L., MacAllister, I., Tunstall, H. and Rossiter, D. J. (2000) 'Local context, retrospective economic evaluations, and voting: the 1997 general election in England and Wales', *Political Behavior*, 22, 121–143.

Johnston, R. J., Pattie, C. J., Dorling, D. F. L., MacAllister, I., Tunstall, H. and Rossiter, D. J. (2001a) 'Housing tenure, local context, scale and voting in England and Wales, 1997', *Electoral Studies*, 20, 195–216.

Johnston, R. J., Pattie, C. J., Dorling, D. F. L., MacAllister, I., Tunstall, H. and Rossiter, D. J. (2001b) 'Social locations, spatial locations and voting at the 1997 British general election: evaluating the sources of Conservative support', *Political Geography*, 20, 85–112.

Johnston, R. J., Pattie, C. J., Dorling, D. F. L. and Rossiter, D. J. (2001a) *From Votes to Seats: The Operation of the UK Electoral System since 1945*. Manchester: Manchester University Press.

Johnston, R. J., Pattie, C. J., Dorling, D. F. L. and Rossiter, D. J. (2001b) 'Distortion magnified: New Labour and the British electoral system, 1950–2001', in C. Rallings et al. (eds), *British Elections and Parties Review* (Volume 12). London: Frank Cass.

Johnston, R. J., Pattie, C. J., MacAllister I., Rossiter, D. J., Dorling, D. F. L. and Tunstall, H. (1997) 'Spatial variations in voter choice: modelling tactical voting at the 1997 general election in Great Britain', *Geographical & Environmental Modelling*, 1, 153–179.

Johnston, R. J., Pattie, C. J. and Rossiter, D. J. (2001) 'He lost…but he won! Electoral bias and George W. Bush's victory in the US Presidential election, 2000', *Representation*, 38, 150–158.

Johnston, R. J., Pattie, C. J. and Russell, A. T. (1993) 'Dealignment, spatial polarisation and economic voting: an exploration of recent trends in British voting behaviour', *European Journal of Political Research*, 23, 67–90.

Jones, K. (1991) 'Specifying and estimating multi-level models for geographical research', *Transactions, Institute of British Geographers*, (NS) 16, 148–159.

Jones, K., Gould, M. I. and Watt, R. (1998) 'Multiple contexts as cross-classified models: the Labour vote at the British general election of 1992', *Geographical Analysis*, 39, 65–93.

Jones, K., Johnston, R. J. and Pattie, C. J. (1992) 'People, places and regions: exploring the use of multi-level modelling in the analysis of electoral data', *British Journal of Political Science*, 22, 343–380.

Karvonen, L. and Kuhnle, S. (eds) (2001) *Party Systems and Voter Alignments Revisited*. London: Routledge.

Key, V. O. (1949) *Southern Politics*. New York: Vintage Books.

Key, V. O. (1955) 'A theory of critical elections', *Journal of Politics*, 17, 3–18.

King, G. (1989) 'Representation through legislative redistricting: a stochastic model', *American Journal of Political Science*, 33, 787–824.

King, G. (1997) *A Solution to the Ecological Inference Problem*. Princeton, NJ: Princeton University Press.

King, G. and Browning, R. X. (1987) 'Democratic representation and partisan bias in Congressional elections', *American Political Science Review*, 81, 1251–1273.

Lennertz, J. (2000) 'Back in their proper place: racial gerrymandering in Georgia', *Political Geography*, 19, 163–188.

Lewis-Beck, M. S. and Stegmaier, M. (2000) 'Economic determinants of electoral outcomes', *Annual Review of Political Science*, 3, 183–219.

Lijphart, A. (1994) *Electoral Systems and Party Systems*. Oxford: Oxford University Press.

Lipset, S. M. and Rokkan, S. E. (1967) 'Cleavage structures, party systems and voter alignments', in S. M. Lipset and S. E. Rokkan (eds), *Party Systems and Voter Alignments*. New York: The Free Press, pp. 3–64.

MacAllister, I., Johnston, R. J., Pattie, C. J., Tunstall, H., Dorling, D. F. L. and Rossiter, D. J. (2001) 'Class dealignment and the neighbourhood effect: Miller revisited', *British Journal of Political Science*, 31, 41–59.

MacAllister, I. and Studlar, D. T. (1992) 'Region and voting in Britain: territorial polarization or artefact?', *American Journal of Political Science*, 36, 168–199.

Marsh, M. (2002) 'Electoral context', *Electoral Studies*, 21, 207–218.

Miller, W. E. and Shanks, J. M. (1996) *The New American Voter*. Cambridge, MA: Harvard University Press.

Miller, W. L. (1977) *Electoral Dynamics*. London: Macmillan.

O'Loughlin, J. (2000) 'Can King's ecological inference method answer a social scientific puzzle: who voted for the Nazi party in Weimar Germany?', *Annals of the Association of American Geographers*, 90, 592–600.

O'Loughlin, J., Ward, M. D., Lofdahl, C. L., Cohen, J. S., Brown, D. S., Reilly, D., Gleditsch, K. S. and Shin, M. (1998) 'The diffusion of democracy, 1946–1994', *Annals of the Association of American Geographers*, 88, 545–574.

Osei-Kwame, P. and Taylor, P. J. (1984) 'A politics of failure: the political geography of Ghanian elections, 1954–1979', *Annals of the Association of American Geographers*, 74, 574–598.

Pattie, C. J., Dorling, D. F. L. and Johnston, R. J. (1997) 'The electoral geography of recession: local economic conditions, public perceptions and the economic vote in the 1992 British general election', *Transactions of Institute of British Geographers*, (NS) 22, 147–161.

Pattie, C. J. and Johnston, R. J. (2000) '"People who talk together vote together": an exploration of contextual effects in Great Britain', *Annals of the Association of American Geographers*, 90, 41–66.

Pattie, C. J. and Johnston, R. J. (2001) 'Routes to party choice: economic evaluations and voting at the 1997 British general election', *European Journal of Political Research*, 39, 373–389.

Pattie, C. J. and Johnston, R. J. (2002) 'Political talk and voting: does it matter to whom one talks?', Environment and planning A, 34: 1113–1135.

Pattie, C. J. and Johnston, R. J. (2003) 'Hanging on the telephone? Doorstep and television campaigning at the 1997 British general election', *British Journal of Political Science*, 33, 303–322.

Pattie, C. J., Johnston, R. J. and Fieldhouse, E. A. (1995) 'Winning the local vote: the effectiveness of constituency campaign spending in Great Britain, 1983–1992', *American Political Science Review*, 89, 969–986.

Pattie, C. J., Whiteley, P., Johnston, R. J. and Seyd, P. (1994) 'Measuring local political effects: Labour Party constituency campaigning at the 1987 general election', *Political Studies*, 42, 469–479.

Rae, D. W. (1967) *The Political Consequences of Electoral Laws*. New Haven, CT: Yale University Press.

Reynolds, D. R. (1969) 'A "friends-and-neighbors" voting model as a spatial interactional model for electoral geography', in K. R. Cox and R. G. Golledge (eds), *Behavioral Problems in Geography*. Northwestern Studies in Geography, 17, Evanston, IL: Northwestern University Press, pp. 81–100.

Reynolds, D. R. (1990) 'Whither electoral geography? A critique', in R. J. Johnston, F. M. Shelley and P. J. Taylor (eds), *Developments in Electoral Geography*. London: Croom Helm, pp. 22–38.

Rossiter, D. J., Johnston, R. J. and Pattie, C. J. (1999) *The Boundary Commissions: Redrawing the UK's Map of Parliamentary Constituencies*. Manchester: Manchester University Press.

Russell, A. T. (1997) 'A question of interaction: using logistic regression to examine geographic effects on British voting behaviour', in C. J. Pattie et al. (eds), *British Elections and Parties Review* (Volume 7). London: Frank Cass, pp. 91–109.

Scarbrough, E. (1984) *Political Ideology and Voting*. Oxford: Clarendon Press.

Shelley, F. M., Archer, J. C., Davidson, F. M. and Brunn, S. D. (1996) *Political Geography of the United States*. New York: Guilford Press.

Shugart, M. S. and Wattenberg, M. P. (eds) (2001) *Mixed-Member Electoral Systems: The Best of Both Worlds?* Oxford: Oxford University Press.

Stoker, L. and Bowers, J. (2002) 'Designing multi-level studies: sampling voters and electoral contexts', *Electoral Studies*, 21, 235–268.

Sui, D. L., Fotheringham, A. S., Anselin, L., O'Loughlin, J. and king, G. (2000) 'Book review forum', *Annals of the Association of American Geographers*, 90, 579–606.

Taagepera, R. and Shugart, M. S. (1989) *Seats and Votes: The Effects and Determinants of Electoral Systems*. New Haven, CT: Yale University Press.

Taylor, P. J. (1978) 'Political geography', *Progress in Human Geography*, 2.

Taylor, P. J. (1986) 'An exploration into world-systems analysis of political parties', *Political Geography Quarterly*, 5, 5–20.

Taylor, P. J. (1990) 'Extending the world of electoral geography', in R. J. Johnston, F. M. Shelley and P. J. Taylor (eds), *Developments in Electoral Geography*. London: Croom Helm, pp. 257–271.

Taylor, P. J. and Flint, C. (2000) *Political Geography: World-Economy, Nation-State and Locality* (Fourth Edition). London: Longman.

Taylor, P. J. and Johnston, R. J. (1979) *Geography of Elections*. London: Penguin.

Thrift, N. J. (1983) 'On the determination of social action in space and time', *Environment and Planning D: Society and Space*, 1, 23–57.

Tufte, E. R. (1973) 'The relationship between seats and votes in two-party systems', *American Political Science Review*, 67, 540–554.

Tunstall, H., Johnston, R. J., Rossiter, D. J., Pattie, C. J., MacAllister, I. and Dorling, D. F. L. (2000) 'Geographical scale, the "feel-good factor" and voting in the 1997 general election in England and Wales', *Transactions, Institute of British Geographers*, (NS) 25, 51–64.

Whiteley, P. and Seyd, P. (1994) 'Local party campaigning and electoral mobilisation in Britain', *Journal of Politics*, 56, 242–252.

Wleizen, C. and Franklin, M. (2002) 'Reinventing election studies', *Electoral Studies*, 21, 157–160.

Wright, G. C. (1977) 'Contextual models of electoral behavior: the Southern Wallace vote', *American Political Science Review*, 71, 497–508.

4 Representation, Law and Redistricting in the United States

Richard L. Morrill

States in which the perceived interests of individuals and of distinct groups of citizens or of distinct regions can be voiced, through free expression and voting, and potentially implemented by an elected assembly, are representative democracies. The quality of representation and the meaningfulness of these democratic institutions are the subject of this chapter. I begin with the theory of representation, and treat alternative forms of representation, within Western democracies. I also try to place the exercise of voting for representatives within the broader exercise of power within such societies. The next section reviews the evolution of representation, including the conflict between territory and group, especially racial, bases for representation. The political-geographic practice of redistricting is discussed, addressing broad issues of fairness of representation. Finally, fundamental issues of representation and governance across geographic scales are raised, and remaining unresolved problems of redistricting practice are reviewed.

Democracy and Representation

Democracies need and enable collective decisions. Voting enables certain citizens – those deemed eligible to participate – to express their will with respect to issues of collective choice (Dixon, 1968; Pitkin, 1967). Except for a few local governments, the size of territories and of populations precludes direct involvement in collective governance, so the most critical voting decisions are for representation. In Western democracies there are at least four fundamental modes of representation:

1 Directly, as individuals;
2 Indirectly, in support of a party or person which expresses a program of governance with which we agree at the time;

3 Indirectly, in support of a representative who shares a sense of belonging to a particular territory;
4 Indirectly, in support of a representative who shares a racial or linguistic heritage, which has probably been historically suppressed.

The most direct kind of participation is to vote on state and local issues of all kinds, but included as well is voting for executives at all levels, from mayors and sheriffs to presidents. The representatives we vote for potentially match the other three forms: for a party and program which is granted responsibility for organizing and operating government for a period of time; to ensure that needs and interests of our territory (whether a community, our city or country, or our legislative or congressional district) are addressed; and to give members of racial or linguistic minorities a fair possibility of choosing persons from their group, if they so wish.

Citizens could decide all matters by direct plebiscite, but those powers are in practice constrained. They could elect all representatives at large by proportional representation, thus meeting the need for representation by party and perhaps by race or other interests as well, but this is a practice usually only for very local governments and, at least in the United States, for the minority of jurisdictions. A territorial basis for representation as well as for governance itself is a fundamental property of many Western democracies, notably in the US and the UK. Representation is in fact dual: partisan and territorial. The tradition of territorial representation dates to the earliest stirrings of a sense of representation, that of named places (Morrill, 1987). Since it is the people of real places that pay taxes to support national and local activities, it is not unreasonable that they should develop a sense of common interest in how that money is spent. In the US, it was a profound sense of territorial interests (that is, cities, at least prior to 1962) not being fairly represented, not just a sense of partisan misrepresentation, which led to the entire 'reapportionment revolution'.

In the US, states are the only fixed units of representation, for the United States Senate. Thus, it is necessary to de-limit representative districts for the House of Representatives, for all state legislatures, and for the large majority of counties, cities, school districts and other units of local government. The goal of districting is to make possible the meaningful and effective participation of voters in electing representatives. Voters need to feel that voting is worthwhile, that their vote matters, that their interests and those of their community will be considered and have some chance of being addressed. And this is why the geography of electoral districts is important. The geographic design of districts is itself a major element in the balance of power. Manipulation is tempting and perhaps inevitable, but if citizens perceive the system as unfair and corrupt, a sense of disenfranchisement and futility develops, in turn leading to reduced participation, reduced willingness to support government, and to reduced quality of representation and of governance.

Obviously there are very disparate interests, even within a district: not only will there be a winner and loser between the likely two major 'parties', one of which may be out of power for very long periods, but lesser interests are never likely to be embraced. This fact leads to the principle of 'virtual' representation, that in a large, diverse country like the United States, representatives from other jurisdictions indirectly address your interests. Alternatively, forms of proportional representation have been developed to increase the chances of minority interests being voiced in one's own territory.

Even the freest societies constrain who may participate in governance, how they may participate, and what aspects of life are subject to collective (or private) decision-making. Societies typically restrict participation to citizens deemed to have a stake, implying both a right and a responsibility. *Who* is deemed worthy of participation has evolved over time, for example, from males with significant property to most adult citizens. This still excludes the dependent youth (under 18 now in the US), aliens (who presumably can participate in another country), and usually, those in prison. *How* they participate includes provisions for discussion and questioning via hearings and town meetings; rules for formal voting, as well as for at what geographic or political levels one may vote; rules determining the degree to which citizens can decide substantive issues by direct vote; and various forms of representation (single member districts, proportional representation).

Individual states within the US vary greatly in the prevalence of direct voting and the ease of circulating initiatives (submitted by groups of voters) and referenda (submitted by legislative bodies to voters; this may include sizeable capital expenditures), and at what levels these may occur. In the US, many states and local governments have initiatives, referenda, votes on bonds, and so on, but the US Federal government does not. In presidential systems, the electorate does vote for President/Vice President (although in the US in a cumbersome and indirect form). In many countries, citizens directly elect local officials (such as mayors), and in fewer countries, regional executives (for example, Governors). In general, in parliamentary systems, higher-level executives (for example, the Prime Minister) are filled from the body of elected representatives, rather than directly. The indirect form is based theoretically on the precept that voters intend to grant a party responsibility for governance – writing laws and executing them – while the direct form is justified on the grounds of a possible need for 'checks and balances' across government, as well as because of the visibility and significance of the executive to the citizenry.

The theoretical basis for direct voting is that citizens are interested and knowledgeable, but the fact that provisions for direct voting diminish at higher levels of government implies an inverse relation between knowledge and interest and the size (area and population) of the jurisdiction. Especially in earlier periods of rural dominance and poor transportation, most citizens might be expected to know local issues but not wider ones. Thus the theoretical bases for indirect participation in governance via elected representatives is threefold:

1   The nation or many regions are too large, and the issues too complex for the typical working citizen to be expected to comprehend, but chosen representatives could, for specific terms, divert their lives to learn these wider concerns, and to articulate the interests of the voters of their 'places' or 'districts';

2   Although not often expressed overtly, especially now, there was also the assumption that the elected representatives were likely to be inherently more knowledgeable;

3   Because people in places had defined interests, a person who 'represents' both the place and those interests could best express such territorial concerns.

The evolution of representative democracy in Europe and the US was overtly territorial – the parishes, communes, counties, towns, cities and other familiar units of settlement and social organization. Indeed, until relatively recently representation was based on these legally and culturally defined places or areas. Yet these places in fact had strongly differing interests related to farming, mining, industry, transportation, class or ethnicity, and representatives found it essential to forge alliances, that became parties, or factions of parties, which espoused the shifting social and economic cleavages of society – individual and collective, liberal and conservative (with respect to social rights and economic policies), core and periphery, city and country – that are the substance of collective choice. Thus, the alternative form of representation would be by interest rather than by territory. This may be termed 'transcendent' representation, implying that identification with an interest community transcends spatial limits (Kelly, 1996; Leib, 1998). A syndicalist form of representation would be based on occupation or industry of the working citizenry, wherever they might live. In a later section, I observe how the power of citizens organized into interest groups may well *de facto* equal that of formal legislative bodies. Or, especially in complex multi-ethnic states, representation could be by ethnic identity. In the United States, racial and ethnic status is privileged to varying degrees, in contradiction to the territorial norm of representation, and indeed in the 1990s congressional and legislative districts were drawn which totally ignored any traditional sense of territoriality, and which were defended on the argument that racial group identification was more salient than territorial community. In as many as 19 countries, some legislative seats may be reserved for numerically small and/or scattered ethnic or racial groups (for example, several groups in India, the Inuit of Canada, or the Aborigines of Australia).

The prevalence of participation in the Internet, together with increasing use of absentee ballots, and the high level of mobility of populations in the richest democracies, has led some to declare that space is dead, and that a territorial basis for representation is no longer defensible. But especially in our complex society we forge so many identities (gender, social issues, economic concerns, lifestyles, work, leisure, as well as race or ethnicity or multiple races) that, in the end, territory remains as the single broadest community basis for representation.

Indeed, in attempting to avoid recognizing a group racial interest (see below) the courts have embraced territoriality or the supremacy of regional community representation.

Systems of Representation

The main distinction in Western democracies is between effectively single member districts – the winner-take-all system – and proportional representation within multi-member districts. In most English-speaking countries, but also in Japan, France and several other smaller countries, the nation's territory is divided into an exhaustive system of districts (for whatever level of governance) from which one representative is elected, usually after party nominees are chosen in a primary election. Because the winner is whoever gets the most votes, no matter how close the result, or how many candidates there may be, the system tends to discourage more than two parties. It also tends to move those two parties toward the ideological center; tends to result in dramatic shifts in party strength; and can encourage a sense of unfairness to the minority parties, from a likely difference between shares of the vote and shares of seats won (Wildgen and Engstrom, 1980). Because of the federal system in the US, and dramatic economic, social and ethnic variation across the nation, and between rural and metropolitan regions, some of these effects are not so severe as in, for example, the United Kingdom. For the US Congress as a whole the difference between the proportion of votes and proportion of seats is typically no more than 2%. However, the average difference at state level is about 10%, with the Democratic advantage in some states counteracted by a Republican advantage in others. Similar to the UK, Democrats in the US tend to get more seats than votes, because of their domination of low turnout inner-city districts. In the UK, small shifts in voter sentiment can lead to large shifts in effective representation (Johnston et al., 2001). There are also variants on single-member systems. The Australian Senate employs an 'alternative vote' system to correct somewhat for electoral bias. France requires a run-off or double ballot to ensure that the winner obtains a majority. Japan elects 511 Diet members from 124 districts; since each voter gets only one vote, much of the bias can be overcome.

The single-member system of voting is defended theoretically on the grounds of effective governance. The argument is that two broad parties are preferable to many specialized ones, and even that disproportionate strength is useful to allow the winning party to follow through on the policies that the majority of voters supported. It is also defended in terms of territoriality – on the idea of a special relationship between a representative and the place that elected him or her. The fact that voters of the losing party may be unable to elect the person of their choice indefinitely, because of the uneven geography of interests, is justified on the grounds of 'virtual' representation, that the interests of such voters are met through districts that routinely return the other party. This tends to be numerically true, but of course contradicts the very basis of territorial representation.

Since winner-take-all systems tend to be dominated by two parties, which gravitate toward the center in order to win power, a sizeable part of the electorate may feel disenfranchised, and often advocate a shift to proportional representation. Proportional representation would tend to benefit any smaller, less mainstream party, major parties in regions where they are demographically weak, and racial, ethnic or any other minority interests. The perceived sense of unfairness and disenfranchisement from the winner-take-all system has led many countries, including most of Asia, continental Europe and Latin America, to adopt varying forms of proportional representation within multi-member districts. The nation may be divided into several regions, from which varying numbers may be elected; these regions may well be traditional legal and cultural sections of the country. Israel and The Netherlands have only one district: the whole country. There are many specific systems, but the principle is that all parties above some threshold level of support (for example, 5% of votes) will win seats proportional to their share of the overall vote. In this way, class, sub-regional, ethnic and other interests may all be represented. Voters usually do not vote for individuals, but for party slates. Electoral district drawing is thus unimportant, but changes in the boundaries of the administrative territories, for example city boundaries or a new regional entity, can influence the balance of power. The main critique of proportional representation is that it fosters too many parties, and too narrow interests, such that it is difficult to forge a government with a mandate and sense of responsibility to enact a coherent program.

A number of countries have chosen a combination of single-member voting and proportional representation called the 'additional member' system. In both Germany and Russia, for example, half the members of high-level legislative bodies are chosen from single-member districts (as in the US and the UK), but the other half are chosen by PR using large multi-member districts. Voting for the latter list makes it possible to 'top up' or correct any imbalances in the proportion of votes versus proportion of seats gained from the single-member voting.

Political Representation and Power in Social Systems

The ideal of representation does not match the reality of power. In theory, according to the universalist principle, individuals (rather than households, groups, interests or areas) participate in governance directly (through initiatives or referendum) and indirectly (through electing representatives). Of course these individuals, because they belong to wider societies, simultaneously express household, group and area interests, reflecting a particularist principle. In practice, moreover, individuals are not equal. Whatever the social system, the individual voting decision, and even more, the decisions or non-decisions of our representatives, are the results of a complex exercise of power. Western democracies are capitalist, which means that private property rights, proprietary and corporate enterprise, and markets dominate and regulate the economy. But this system has survived centuries of challenge, success, and failure by adaptation, and the result is a surprisingly mixed system, containing different combinations of collective, re-distributive, and planned decision-making.

Power to define the balance of power in society comes not only from the individual voter, but from holders of wealth; from the buyers of labor (both corporate and government enterprise); from thousands of interest groups across a stupendous spectrum of thought; from experts, including the academy; and from the media, which tends to control access to the information upon which decisions are made. This means that the electoral system and issues of redistricting and its fairness may not be as important as some believe. The intelligent citizen is quite aware of this unequal power, and recognizes that his or her influence will be enhanced by participation in interest groups, by contributing to the media, and by acting collectively with others to influence collective decisions and the voting behavior of elected representatives. In addition, we know that decisions of most vital importance to our neighborhood and our own livelihood are likely to be made at a distant, higher level of governance, or in the private sector, despite the rhetoric of local government and a territorial basis for representation.

Changing Meanings of Representation and the Practice of Redistricting

Extending the Franchise

Discontented as we may be with the unfairness of electoral systems, it was far worse in the past. The franchise was gradually extended over the centuries, and the structure of electoral districts came to reflect more closely the sentiment of voters. Those in power have always been loath to share governance and risk the erosion of their power and wealth. Although some degree of participation was extended in the later Middle Ages to larger landowners, and perhaps a little more broadly in Scandinavia than in Britain, the rise of capitalism, mercantilism, industry and cities, and the beginning of a middle class were critical. The complexity and cost of government, defence and warfare increased, and urban production and revenues were able to act as leverage for urban commercial and industrial participation and raise the role and power of legislative bodies.

The principle that only men with substantial property had a sufficient stake to vote also prevailed in the early American Republic. The extension of the franchise subsequently followed on from periods of war. After the war of 1812, veterans with less property were rewarded with the vote, leading to the first populist Jacksonian shift of government. Following the US Civil War, the vote was extended to all adult males, regardless of property and, theoretically, to freed slaves (the 15th Amendment). That right was short lived, as the white power structure soon erected Jim Crow barriers to effectively stop black participation for another century. Women played a significant role in the First World War, and this was instrumental in gaining women the right to vote. A few states in the West had earlier allowed women to vote, in similar recognition of their greater economic role on the frontier. The effective extension of votes to African Americans in the South was more complex, beginning with Truman's integration of the Armed Forces in 1949 and the subsequent role of African Americans in the Korean War. Even that was not enough – it took race riots and church burnings, threatening wealth and property, to bring about the Civil Rights and Voting Rights Acts. The last extension of the franchise, to 18 year olds, rewarded their involvement in the Vietnam War – they were able to die at 18, but not vote for those who sent them there.

Malapportionment: Who counts?

Malapportionment means unequal distributions of population between districts. A citizen of a district containing 1,000 voters will clearly have twice the 'electoral power' of those in a district with 2,000 voters. This would appear to violate a sense of fairness in the social contract. Complaints were made over the centuries about the unequal size of constituencies for the British House of Commons, and periodically some of the worst abuses, such as the 'rotten boroughs' of the eighteenth century, were addressed. In the United States, voters of very populous districts complained for over a century about such unfairness, but unequal population was successfully defended on two main grounds:

1 Since states had two US Senators each, whatever the state population, state senates should logically follow the same principle;
2 For the House of Representatives, the perceived greater needs of sparsely populated rural areas, including problems of communication.

This latter principle still carries legal standing in Canada and the UK, although acceptable population variation has been narrowed. In 1960, disparities in the US were as much as 400:1 (for example, for state senate districts in California). Importantly, in the US, the system of Federal Courts, and ultimately the US Supreme Court, have the final word in matters of redistricting. In 1946 (in *Colgrove vs. Green*), a population discrepancy of 9:1 was viewed as acceptable. Not until the US population was majority metropolitan (in 1960) did the sense of the unfairness of the system of districts, in which rural and small-town America still dominated legislatures and the US Congress, finally compel US Supreme Court recognition that there was a constitutional issue at stake (in *Baker vs. Carr*, 1962). In *Reynolds vs. Sims* (1964), the 'one person, one vote' principle was established – that 'people, not trees, vote' – leading to the expectation, after hundreds of additional court actions, of reasonable population equality for districts at all levels of government. For Congress, rather extreme equality was required (*Kirkpatrick vs. Presler*, 1969; *Karcher vs. Daggett*, 1983). Legislatures, counties, and cities were given more leeway, perhaps a spread

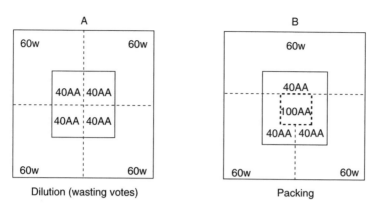

Figure 4.1   Racial gerrymandering (see Morrill, 1981)

W = White voters
AA = African American voters

of +/–10%, if local government boundaries were respected (*Connor vs. Finch*, 1976; *Mahan vs. Howell*, 1973). The analogy between state Senates and the US Senate was also struck down. Yet the meaning of population is still unsettling in the US electoral system. All resident persons, whether citizens or not, of whatever age, are counted, but citizens abroad are excluded. (Utah failed to get another seat in Congress for this reason, but, on the other hand, Utah adults have greater electoral power, because the high proportion of children there awards the state more seats). In the UK, the more intuitively reasonable criterion used is an equal number of potential voters. This was used in Hawaii, but was disallowed when Hawaii became a state. Excessive precision is also undermined by recognition that populations have already changed by the time redistricting is carried out and elections held, and that many people have multiple residences. This has not prevented many states going to heroic and even absurd lengths, by using block data, to get within one or two persons of 'equality'.

Gerrymandering

The next step of the 'reapportionment revolution' addresses gerrymandering, that is, the deliberate manipulation of territory. The term 'gerrymandering' immortalizes the name of the Massachusetts Governor, and the cartoonist, Tisdale, who in 1812 likened the shape of one manipulated district to a salamander. The purpose of gerrymandering is to help 'your side' and to hurt the 'other side', whether this refers to a political party, a racial or ethnic group, kinds of territory (urban, rural, suburban, rich, poor) or individuals (incumbents or challengers), and thereby to obtain a higher share of seats than of votes. There are various means of gerrymandering electoral districts (see Figure 4.1):

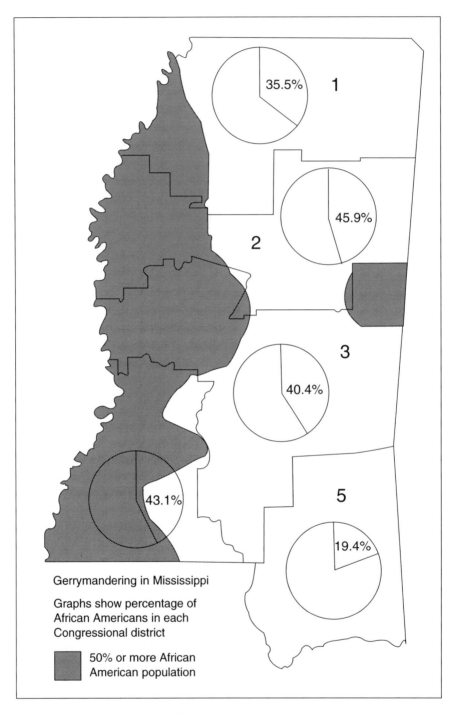

Gerrymandering in Mississippi

Graphs show percentage of
African Americans in each
Congressional district

50% or more African
American population

Figure 4.2   Gerrymandering in Mississippi

1   Splitting or dilution of the concentrations of the other side, so as to leave them a minority in as many districts as possible;
2   Packing or concentrating the other side in as few districts as possible, so that many of their votes are 'wasted', while also creating many districts with moderate margins for your side;
3   Placing incumbents of the other side in the same revised districts;
4   Creating multi-member districts with your party in the majority.

While gerrymandering often does take the form of extremely irregular districts, subtler gerrymandering can appear reasonably compact, while using any one of these four tools. Changes to the congressional districts of Mississippi illustrates the transition from traditional communities to a discriminatory gerrymander against black voters, by splitting the North to South trending community via East-West districts (see Figure 4.2).

The courts have both condemned and endorsed gerrymandering, but only with respect to discrimination against a racial or linguistic minority (O'Loughlin, 1982). How can this be? Early court involvement was cautious, in terms of the 14th Amendment (equal protection) and the 15th Amendment (extending the franchise to freed slaves). The critical decisions were made on the basis of the Voting Rights Acts (1965, and subsequently amended), which outlawed dilution, disallowed retrogression, and required 'preclearance' of plans by the Department of Justice in states with a history of racial discrimination.

The question of racial fairness had been declared judiciable in *Fortson vs. Dorsey* (1965), but discrimination proved hard to prove. *White vs. Regester* (1973) ruled that a plan could be evaluated on the basis of its effects (or lack of), but *Mobile vs. Bolden* (1980) ruled that an intent to discriminate had to be proved (O'Loughlin and Taylor, 1982). The most definitive court decision was *Thornburg vs. Gingles* (1986), yet it has by no means proven to be a clear guideline for race-conscious redistricting. *Gingles* established the basic standard, that race can be one but not the controlling factor in drawing a district, and that it is reasonable to design so-called 'majority minority' districts, if there is a history of block voting, and if the minority population is geographically insular, so that reasonable compactness can be maintained. In effect, only segregated racial or ethnic communities constitute an acceptable basis for representation. The basic problem with race-aware redistricting was the risk of packing in the large urban ghettoes, and the difficulty of assembling sufficient minority numbers in rural and small-city parts of the South.

Despite *Gingles*, the Department of Justice, many state legislatures, and local courts began the 1990 round of redistricting with the expectation that a plan should strive for proportionality, even if heroic gerrymandering were required (Pildes and Niemi, 1993). This led to the creation of well-publicized irregularities in several redistricting plans. Perhaps the best known, North Carolina's original 12th district, connected black communities from Charlotte and Gastonia in the southwest to Durham in the northeast by strips of Interstate

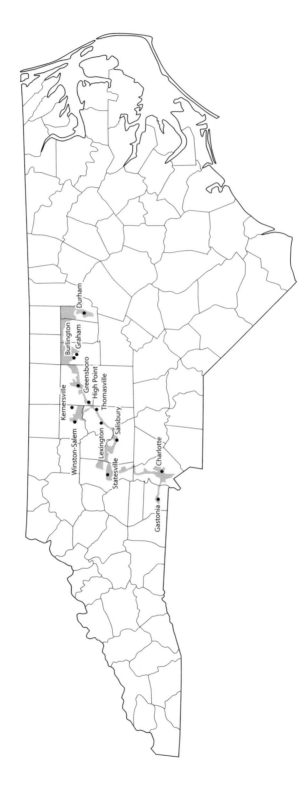

Figure 4.3 The 12th Congressional District of North Carolina, 1992

85 as narrow as the freeway right-of-way (see Figure 4.3). So complex were the boundaries that the North Carolina 12th District required 30 pages in the official Congressional District Atlas. This district was not a great deal worse in compactness measures than the North Carolina 1st District and others in Florida (3rd), six districts in Texas, New York (12th), Louisiana (4th), Georgia (11th) and Illinois (4th). Although the gerrymandered districts did elect a record number of minority members in the 1992 elections (Ingalls et al., 1997), this was at the expense of the loss of even more liberal white Democrats, suggesting that the goal of maximizing minority representation was not necessarily the most effective way of furthering minority interests (Canon, 1999; see also Leib, 1998; Webster, 1993).[1]

Voters in several states filed actions against plans, alleging that race was the controlling criterion rather than just one criterion, as had been established in *Gingles*. In the majority of these cases, the Supreme Court has held plans to be unconstitutional racial gerrymanders, and required less draconian and more regular plans. *Shaw vs. Reno* (1993) held that the North Carolina districting plan warranted 'scrutiny'. *Shaw vs. Hunt* (1996), also called *Shaw II*, found the 12th district to be an unconstitutional racial gerrymander, and that the outrageous irregularity of the district violated principles of territorial community representation. Not until *Hunt vs. Cromartie* (1999) were the North Carolina congressional districts resolved. Louisiana's convoluted 4th district, or 'Mark of Zorro', was struck down in *Hays vs. Louisiana* (1993s and 1996); Georgia's 11th in *Miller vs. Johnson* (1994 and 1995); and Texas' 18th, 29th and 30th in *Bush vs. Vera* and *Vera vs. Bush* (both 1996). Yet the infamous Illinois 4th, the 'Earmuff district', and New York's 12th, the 'Bullwinkle' district, were found to be acceptable. As a result of the necessary revisions, which created far more regular districts, several incumbents had to move to adjacent districts where more of their original constituents now resided. Although no longer 'majority minority' districts, almost all the districts re-elected minority incumbents, even though the districts were quite different, illustrating the great power of incumbency in the US.

As an alternative to racial gerrymandering, Lani Guinier (1995) has proposed cumulative voting in a few multi-member congressional districts as an alternative to racial gerrymandering, and alternative plans were proposed for North Carolina and other states. However, citizens and the courts seem unwilling so far to take this logical next step, at least for congressional districts. At the local level, courts have approved and even required the use of proportional representation in multi-member districts as a means to enable minority representation. For example, cumulative or limited voting is used in selected Alabama, North Carolina, Texas and other state county boards of commissioners, small-city councils, and school district boards.

Proportional representation (PR) is not absent in the US and other nations with dominantly single-member systems (Richie, 1998). Dozens of local governments have adopted a variety of PR systems, and the courts or local jurisdictions have intentionally adopted PR to deal with the issue of fair representation of racial minorities. For example, county boards of commissioners, or school districts, or city councils may use systems like cumulative voting (used in Texas) or limited voting (used mainly in Alabama and North Carolina) to make it more likely for minority members to be elected. In cumulative voting, for example, for three representatives, voters can concentrate their votes on two or even one candidate. In limited voting, for the same three representatives, voters may only be able to choose two, with a similar effect. Limited voting can provide effective minority representation (Arrington and Ingalls, 1998; Brischetto and Engstrom, 1998).

This leaves unresolved two aspects of gerrymandering and representational fairness – those of party and of territory. There is no academic or judicial consensus to invalidate manipulation against a political party or recognized place or territorial community (Baker, 1986). Compactness and community of interest have been recognized as 'desirable' but not as essential. Courts have ruled against the recognition of any kind of community except race or language (*United Jewish Organizations of Williamsburg vs. Carey*, 1977), and thus far, against evaluation of political gerrymandering. In *Bandemer vs. Davis* (1984), the US District Court ruled an Indiana legislative plan unconstitutional on grounds of diluting Democratic voting strength, but this was subsequently overturned by the Supreme Court (*Davis vs. Bandemer*, 1986). In *Badham vs. Eu* (1987), an egregious and trumpeted political gerrymander of California congressional districts was similarly upheld. Monmonier (2001) highlights the inconsistency of the courts in striking down racial gerrymanders, in part via explicit reference to the inexcusable irregularity of their boundaries and by invoking the importance of geographic community, yet totally failing to be concerned with similar convoluted districts in the same and other states, which are evidently acceptable because they were 'political' not 'racial'.

As of the 2000 round of redistricting, the courts have not clearly articulated what is or is not acceptable as to racial fairness (Forest, 2001). There will be a temptation to reduce the weighting of race in the process, but the courts are not likely to permit much, if any, retrogression. While the plans may not exhibit as much extreme irregularity as in the disallowed plans, the availability of GIS (Geographical Information Systems) and partisan greed will probably continue to erode the correspondence of districts with recognized geographic communities, despite court rhetoric. We should not forget that extreme irregularity is less necessary to perpetrate partisan gerrymanders than racial ones, especially given the availability of block data (Gudgin and Taylor, 1978).

Redistricting Practice

Does it matter who redistricts? Whether the plans are drawn by a partisan legislature, a bipartisan one, a bipartisan or non-partisan commission, or by the courts? Canada has had boundary commissions since 1964, and recently reduced allowable population deviations to +/-5%. The UK has redistricted four times, since boundary commissions were established in 1948 (Rossiter et al., 1999). In the US, redistricting is required after each decennial census, for virtually all levels of government, from local school districts and town councils to the US Congress (Niemi and Deegan, 1978). Although individual states and localities may have additional criteria, the controlling criterion is 'appropriate equality of population', and a prohibition against discrimination on the basis of race or minority language. The appropriate degree of population equality required in redistricting in the US is under 1% for congressional districts, but may be as loose as +/–10% for districts of local governments or even state legislatures. The next most common official criteria are avoidance of splitting local government units like counties or cities, and perhaps a phrase about preserving communities of interest. Until the recent emphasis on population equality it was customary for districts to consist of whole counties and cities. But the most important considerations – incumbent protection and partisan considerations – are of course not mentioned. Incumbent protection may be expressed more positively as changing the districts as little as possible for the sake of voter continuity. If possible, partisan redistricters will manipulate the drawing to their advantage. A few studies have evaluated congressional and/or legislative plans with respect to such criteria as racial fairness, both in the sense of avoiding discrimination and in the sense of trying to achieve proportionality; political fairness, in the sense of whether the party share of seats match the party share of votes, at least over a set of elections; whether city and county borders are maintained; whether communities of interest are observed; whether districts are reasonably compact, and extreme irregularity avoided; and whether the process is open and fair.[2]

State legislatures in the United States not only decide on the redistricting of themselves, but also of congressional districts, and usually set the rules for the redistricting of lower levels of government. In the majority of states, legislatures redraw themselves and congressional districts, but in 17 states, they or the voters have wisely turned the process over to commissions of varying degrees of independence. In six states, commissions are only for back up if the legislature fails; legislatures are redistricted in seven states, and both legislatures and congressional seats in four states. Iowa is the only state that excludes political considerations completely. In an unpredictable number of jurisdictions from local governments to states, courts may intervene, sometimes with respect to population deviations, more commonly as to racial fairness. So important are the two dominant parties in the US that commissions usually take a bipartisan form. Commissions may rely on a 'redistricting master', but will also draw up plans themselves or submit their party's plans and look at plans submitted by any number of interest groups. As the 2000 Presidential election showed, the US electorate is quite evenly divided on partisan and on ideological dimensions. As a result, only a minority of states is dominated by one party controlling both houses of the legislature and the governorship. In such cases highly partisan districting may be expected, yet heroic manipulations may not be necessary. In states split between parties, some degree of bipartisan districting is required. In practice, each party in both houses is likely to develop and offer somewhat distinct plans, requiring complex negotiations, the main result of which is, of course, incumbent protection.

In the 1980s, redistricting was dominated by population equality and avoidance of racial dilution or packing; in the early 1990s by aggressive affirmative racial gerrymandering. In the 1980s, irrespective of who was responsible for plans that ultimately survived US Department of Justice review and court scrutiny, all plans greatly improved racial minority representation, with court plans a little more likely to sacrifice other criteria in favor of racial equity. Bipartisan drawn plans tended to achieve reasonable political fairness while preserving incumbents, also tended to create safe rather than balanced or swing districts, and at least paid some attention to city and county integrity and communities of interest. Partisan-drawn plans were predictably biased and more blatantly irregular. California Democrats were especially successful at congressional and legislative plans that awarded far more seats than their share of the votes warranted. In some states partisan districters were overconfident and created too many small-margin districts, resulting in unexpected losses from a small shift in voter sentiment. Partisan plans were poor at city, or county, or community integrity, and both partisan and bipartisan redistricting minimized meaningful public participation. Commission-drawn plans generally tried to achieve political fairness, with a risk of greater volatility from creating too many swing districts. Ironically, a good goal of political fairness could result in greater unfairness in a winner-take-all single-member system. Commission plans were superior at maintaining city and county integrity, and usually included far more public participation. The circumstances of court intervention were so varied that it is hard to generalize.

In the 1990s in many states affirmative racial gerrymandering prevailed; that is, maximizing minority representation was the controlling criterion at the expense of any sense of a geographically defined community. This was encouraged by the Republican Department of Justice in the Reagan administration, with the effect of electing many more minorities but at the expense of the loss of many liberal inner-city Democrats. The overriding of the more traditional territorial sense of community extended beyond the irregular minority districts to ripple across a much larger set of neighboring white districts. In the 1990s, too, partisan plans were more politically unfair than bipartisan or commission plans, but as there were more commission plans, the average level of fairness was probably a little higher than in the 1980s. Whoever redistricted, the 1990s plans tended to divide more cities and counties and paid less attention than ever to communities of interest, because of the greater concern for minority representation and above all because of the availability of data at the block level and of GIS programs permitting the easy manipulation of vast amounts of data. So, together, race, block census data and GIS, and partisan greed resulted in the deterioration of the quality of systems of districts. Finally, in the 1990s, public participation was far more prevalent than in the 1980s, but was usually more pro forma than effective. Monmonier (2001) demonstrates how maps were used as propaganda tools in the hearings process, which he aptly characterized as 'pseudo-populist'. For example, commissions or legislatures might publicize a willingness to 'Help us draw the lines!' but subsequently ignore such contributions.

In general, for the 1980s and 1990s, plans for legislatures and local governments could be evaluated as less irregular and more responsible to traditional communities than plans for congressional districts. Also, on average, larger states exhibited greater manipulation and irregularity, unfairness and failure to respect communities than did smaller states, perhaps because they tended to be more evenly balanced between parties, so the stakes were higher.

Census Data, Maps and GIS in Redistricting

Whoever redistricts, redistricting relies on census population data, maps, GIS programs, and, usually, partisan election returns, incumbent preferences, and participation of interest groups (Clark, 1992). In the 1970s and 1980s maps were poor, population block data were only available for cities, and GIS packages were primitive, and relatively fewer plans were drawn up and compared. By the 1990s, population data became available nationwide at the block level, together with nationwide geography line and polygon files, the TIGER files, and sophisticated GIS programs were developed for interactive districting, as well as the easy capability of manipulating enormous data bases of population and political voting data. In addition, these capabilities were available not only to the parties and the legislatures, but also to the press and to interest groups (Raburn and Leib, 1994).

Although rather attractive size-constrained location-allocation algorithms are available, which incorporate the twin criteria of equal population and compactness – in the form of travel minimization – such models are rarely, if ever, used. I have found that models are especially good at developing a set of alternative plans, which could then be adjusted by additional criteria or local knowledge. However, such models are impractically slow and perform poorly given the block data that are preferred by redistricters/gerrymanderers (there are over 7 million blocks for the 2000 census). Tests of optimizing models by Altman (1998) suggest that while they can limit gerrymandering irregularity, they do not necessarily help with political fairness. Cirincione et al. (2000) used similar optimizing models to show that the South Carolina 6th Congressional District, which was not challenged, was in fact a racial gerrymander.

Overwhelmingly, redistricting now relies on interactive GIS packages which allows one to draw boundaries between districts on the screen, where the base map is the TIGER block map, while an inset or another screen could continuously total the populations. Block groups or blocks could be easily switched with the demographic and even political effects updated. Monmonier (2001) dubs the effect of such manipulations, if aimed at an individual, as 'carto-assassination'. I worked on plans in both California and in Illinois. I will admit it was fun, but my efforts were overruled in both cases because I refused to follow the overriding rule: string those minority blocks together to create more minority districts, whatever the cost to other criteria. In later court decisions, notably *Bush vs. Vera* (1996), heavy reliance on block data became suspect as evidence of a likely racial gerrymander. Still, this is unlikely to result in any retreat from the excessive use of block data for partisan gerrymandering.

Future Issues for Electoral Systems and Legitimate Governance

Across many Western democracies there is widespread dissatisfaction with the electoral system and the quality of representation. In the US, continuing and unresolved issues are:

1   Effective and fair representation for races and linguistic minorities;
2   Partisan gerrymandering and electoral bias;
3   The sense of non-representation for those out of the mainstream consensus;
4   The displacement of power to higher, more distant levels of government;
5   The corrupting effects of campaign finance;
6   And questions surrounding the legitimacy of the 2000 Presidential election.

Representation of Minorities

As a new round of redistricting begins following the release of 2000 population census numbers, uncertainty reigns surrounding the necessary or acceptable degree of attention to race and ethnicity. While the Voting Rights Act remains in place, including provisions of no net retrogression in representation, recent US Supreme Court cases, let alone the new Republican administration, reinforce an acceptance that affirmative gerrymandering, or maximization of minority representation, is proscribed. At most, race can be one serious consideration in the drawing of districts (Epstein and O'Halloran, 1999). Widespread public acceptance of the retreat from race-conscious districting reflects a similar shift with respect to education and employment. The fact that incumbent minorities were generally re-elected, even as their districts were redrawn and were no longer 'majority minority', has led to an undoubtedly false perception that racism in voting is dead. Lennertz (2000) fears that the new emphasis on territorial representation will perpetuate inequalities based on the history of where minorities and the poor could live. Webster (2000) shows how too narrow an emphasis on geographic community would, in a state with a history of severe racial discrimination, like Alabama, result in little or no representation. Yet he observes that successful racial gerrymandering in Alabama did have the perverse effect of polarizing the electorate and severely weakening the Democratic Party and white liberals.

The 2000 census also revealed a slight but widespread decline in levels of segregation, especially among middle- and upper-class minorities most likely to vote, which makes the drawing of race-conscious districts even more difficult. There is a small hope that legislatures may consider multi-member districts with some form of proportional representation (cumulative voting, single transferable vote), at least for local government. I doubt it will be tried for legislative or congressional districts, although, as in North Carolina, it would be the most effective means to enable minority representation while meeting traditional criteria, including community of interest. Nevertheless, the best guess is that most states will be cautious and strive to maintain the level of representation that has been achieved and therefore is perceived as both fair and deserved (Lublin, 1997). This will require that race continue to be a big part of the redistricting process, and that even if 'majority minority' districts cannot be designed, effective shares of 40% or more can often be achieved.

A related issue is that of census adjustment. The US census of population is a massive undertaking, and valiant efforts to count everyone resulted in higher than expected counts. However, there are errors – up to 2% were counted twice, and up to 4% were missed, or intentionally avoided being enumerated (Choldin, 1994). The main reason for overcount is that college students may be counted at the college residence and by their parents at their home; also many people, not just elderly, forgot they were counted and are counted again. Those not counted include probably millions of people who wish to remain anonymous – illegal aliens, criminals, but also many homeless, or travelers. The fact that the undercounted are more likely to be racial minorities or the less well off creates a political incentive for Democrats to favor possible adjustment of census numbers to increase representation from inner cities, Indian reservations, and so on, and for Republicans to resist such adjustment – even though, ironically, Texas and Florida are states that would gain from adjustment. If the census had been adjusted in 1990, California would have gained at least another congressional seat.

The need for or justification for adjustment derives from two sources: demographic analysis which proceeds from vital records of births and deaths, and compares actual and expected numbers of categories or strata of people; and from the detailed post enumeration sample resurvey of about 3% of the population. Typically, census statisticians favor adjustment because they believe they can improve the overall numbers. In the adjustment process, households and people have to be removed and more have to be added, based on calculated shares of strata of people. So long as these are aggregations for fairly large areas, adjustment will improve figures. Indeed, the census is already 'adjusted', in the sense that characteristics of people who fail to answer some questions are assigned imputed answers on the basis of 'similar' people who did answer. The difficulty, and why adjustment is discussed here, is that because redistricting relies so heavily on block level data, any adjustment must go down to the block level, where almost everyone agrees the reliability of data will be less than the actual count, but that the errors will be compensated when aggregating over hundreds of blocks. As a geographer and statistician, I am very resistant to publishing such adjusted data at the block level, and I suspect the courts and the new administration will decide not to adjust census figures. In the end, for 2000, a Democratic vs. Republican fight was avoided when the Census Bureau's own review panel concluded that adjustment would not offer a sufficiently more accurate count, partly because the enumeration proved better than expected.

Partisan Gerrymandering

The best expectation with respect to partisan gerrymandering is no improvement whatsoever. Both parties intend to and will try to draw systems of districts that will hurt the other party and help their own. And the courts will continue to refuse to address the issue. Electoral bias and perceived unfairness will continue to prevail after the current round of redistricting in the United States, and, probably, in most other nations with winner-take-all systems. However, several factors will lessen the degree of bias and unfairness in the United States electoral system. First, 17 states make some use of bipartisan redistricting commissions, and their record is generally recognized as superior to that of legislatures. Second, an even larger number of states typically have divided political control between the legislature and the Governor, or between houses of the legislature, thereby forcing balance if not quality. Third, is the historic tendency to protect incumbents. Fourth, there is the weakness of partisan allegiance, such that voters simultaneously elect Republicans and Democrats for similar offices. Rush (2000) found that as a result of district realignment, large numbers of voters changed their party to match the new district's political culture. This is in contrast to the United Kingdom, where a fairly small shift in the allegiance of marginal voters can lead to a massive shift in elected representatives (Johnson and Pattie, 2001). Finally, the fact that so many interest groups will have the sophistication to review redistricting plans and to draw up their own alternatives may force the process to be more public and may constrain the worst practices and encourage even more court challenges.

Disenfranchisement and Non-representation

The electorates of major Western democracies are amazingly balanced between an ever-changing notion of 'conservative' and 'liberal', as the 2000 Presidential election starkly reveals. Of course the positions of the US Democratic and Republican parties are centrist with reference to the full spectrum of political ideology. The third party candidacies of Buchanan and especially of Nader are a reminder that millions are prepared to vote for their more divergent beliefs. This is not a powerless act – Nader succeeded only too well in electing Bush by siphoning enough votes from Gore in New Hampshire and in Florida. It is true that the two major parties, which legally control the machinery of elections and of governance, do not reflect very well the complexity of the real dimensions of even mainstream liberalism and conservatism. Americans seem distinctly divided along liberal and conservative dimensions on both social and economic issues. As the Republican Party shifted toward a socially conservative agenda, many educated and socially tolerant Republicans, heirs of the historic party, have become independents or even Democrats. And as the Democratic Party embraced social liberalism and environmentalism, it too downplayed its traditional economic populism, and accepted the extraordinary increase in inequality since 1980. Thus, it is the less educated, working-class traditional Democrats, to some degree orphans of the 'new information' economy, that feel the most disenfranchised. In the 2000 election, many deserted the Democrats, especially in much of non-metropolitan America, but apparently stayed with the party in large metropolitan areas, but possibly with decreased turnout, even as social liberal constituencies – racial, sexual minorities, and especially women – showed increased participation and Democratic support.

Serious electoral reform, in the sense of even limited forms of proportional representation, is not going to occur in the United States, with the possible exception of non-partisan councils in a few major leftish metropolitan cities (Amy, 1997). The two major parties, indeed, are terrified at the increase in independent voter rolls and, in a recent Supreme Court case, succeeded in cementing their power through eliminating the open primary (all primary candidates on one ballot, no party affiliation required). In consequence, the recourse of those with a sense of non-representation is two-fold: to migrate to a place or region where one's more extreme values (left or right) may have a chance of electoral enactment; or to participate in the other agencies of political power, such as interest group mobilization. The first is perhaps a main advantage of a federal system of governance over a more centralized one; the latter is in recognition of the multiple ways to exercise power.

Displacement of Power to Higher Levels of Government

Despite the strong tradition of local government, which in a sense is the very basis for the system of territorial representation in the US, and despite the formal division of powers embodied in the federal system, the realities of power and the complexities of modern urban, industrial life create great pressures for the displacement of power to higher levels of government. This in turn contributes to an erosion of trust in government and to resentment across geographic scales and regions. The reasons for the upward shift of decision-making power seem compelling: the highly interdependent metropolitan regional economy, in which labor markets, the commuting system, and air and water quality inevitably encompass hundreds of jurisdictions, therefore requiring 'regional' solutions and institutions. The greater taxing power of higher levels endows the latter with 'carrot and stick' incentive and ability to control lower levels. The perception of a problem at a high level (for example, school achievement, environmental degradation at a state level) results in highly unequal impacts of mandated solutions at the local level (for example, closures of mines, logging, fisheries). Finally, the very great power of interest groups mobilized at a national scale is purposely aimed at legislation at the highest levels.

Thus, even if local areas, and even states, vote for or against a policy or program or development, the other side may well prevail at a higher level of governance, including the courts. This has led to a profound sense of futility and disenfranchisement for local governance, as meaningful control over the local environment has been lost. Jurisdictions within metropolitan regions have lost most of their autonomy to region-wide bureaucracies and planning entities, degrading the value of representation. Even more starkly, rural, small-city and even small metropolitan America has become highly resentful of increased metropolitan concentration of wealth and power. Indeed, the 2000 election may be read as a narrow victory of non-metropolitan America, allied with metropolitan conservatives, against the cultural, economic and perceived excess power of metropolitan liberal elites. But it will not change the reality of the concentration of power.

The Corrupting Effect of Campaign Finance

Given the closeness of the 2000 US election, cynics could well claim that the Republicans bought victory on the basis of superior campaign funding. The amounts spent to influence the voting decision were truly stupendous, and it worked, for wealthy Democrats as well as Republicans. The money comes overwhelmingly from three groups: wealthy individuals, corporations and interest groups. While there are liberal and even leftward wealthy individuals and interest groups, the overall bias of contributions is in a more conservative direction, especially with respect to economic issues. The geographic location of campaign contributions is also extremely concentrated in the wealthy elite areas of major metropolitan areas – this is as true of Democrats as of Republicans – and helps explain the feelings of non-representation of the less affluent. The ideal of 'equal representation' is eroded in four ways: the wealthy voter is contacted more and is more likely to vote; the higher share of conservative messages, on balance, shifts the marginal voter rightward; the elected officials understand who got them elected, and are frequently reminded of that by the major donors; and wealthy individuals and interest groups continue to influence the votes of the officials they were instrumental in electing.

There is widespread cynicism and resentment against the excess influence of the big contributors. This helps explain the popularity of John McCain in the 2000 Presidential campaign. The best expectation is for at least a modest constraint on the magnitude of expenditures, but such limited reform will only marginally affect the electoral and decision-making process.

The Legitimacy of the United States 2000 Presidential Election

The fact that Al Gore won the plurality of the vote in 2000, exceeding George Bush by some 550,000 votes out of over 100 million, is probably the statistic that most intuitively de-legitimizes Bush's presidency, at least in the minds of more than 50 million Gore voters. But there are two subtler aspects that are probably more disquieting to democracy in America. After all, the Electoral College is constitutional, and even if the votes of states had been apportioned proportionally, as in Maine and Nebraska, Bush would have won, simply because he carried far more of the smaller states, thereby receiving the two electors corresponding to their two Senators.

From a probabilistic or statistical point of view, the election was a tie, because in reality, there were probably 2–3 million uncertain votes nationwide, beyond the chaos in Florida. Errors associated with ballot counting have proven to be much greater than most might have believed, but reforms being currently considered are promising. The greater longer-run problem may be the serious weakening of the impartiality and thus the legitimacy of the US Supreme Court, even as it seems more willing to intervene, as a reading of the Court's opinion in this case clearly supports a conclusion that Bush was appointed on the basis of the ideological positions of the Justices.

Conclusion

In the US, elected district representatives carry out most governance. Citizens need to feel that voting is worthwhile, that their vote matters. If the system of representation and the delineation of districts are perceived as unfair, or of creating a sense of disenfranchisement and futility, then democracy is undermined. To voters and their representatives, districts are not merely a passing convenience for the holding of elections, but entities with which they identify. But people have multiple identities: to place, to party and to many other interests. Voters choose representatives, usually of parties, in order to entrust them with governance. In the long term, across a set of elections, a fair and effective democracy insists on a correspondence between the share of votes received and of seats won. The winner-take-all single-member electoral system inherently threatens half or more of voters with imperfect representation. Beyond that, gerrymandering that discriminates for or against racial or ethnic groups, parties, places or classes, further erodes the sense of fair representation. Gerrymandering, made ever easier by GIS, weakens the historic territoriality of representation. It would help if states were to pass laws requiring specific justification for descending to block level data. Gerrymandering obviously works: otherwise why would it be pursued with such fervor and defended with such passion?

The citizen is both an individual rational voter (universalistic) and a member of communities (particularistic). But which communities? The Voting Rights Act asserted that a racial identity is to be protected, but the courts have retreated from that in favor of the traditional regional community, at least rhetorically. Many students of politics, including me, are torn between the competing values of place and interest, of individual choice and group interest, and have come to believe that the only way to preserve the value of place-as-community is to shift to semi-proportional representation, in smallish (2 or 3) multi-member districts, that also simultaneously meet the value of interest-as-community (Morrill, 1999; Webster, 2000). Alternatively, experiments with weighted voting, allowing for unequal populations of districts that preserved known territories, could be explored. Either will be very difficult to achieve. Courts have failed to recognize that it may be impossible to simultaneously retain single-member districts, a commitment to fair political representation, and to recognize racial identity as a legitimate community of interest along with the fundamental regional community. Short of that, the courts have already retreated from race as the only reserved community for representation. It is time, then, to increase the pressure for condemning the cynical use of GIS and gerrymandering to sabotage political fairness, and to restore a sense of the meaningfulness of districts.

Notes

1 See Forest (1995, 2001), Lennertz (2000), Webster (2000) and Grofman et al. (1998) for excellent discussions of the dilemma of attempting to map the geography of racial identity into 'cognizeable' electoral districts, and of the risks to racial fairness from the possible overemphasis on geographic community.
2 See Morrill (1981), Grofman (1982, 1990) and Monmonier (2001) for more detailed discussion of districting criteria, including measures of compactness.

References

Altman, M. (1998) 'Modeling the effect of mandatory district compactness on partisan gerrymanders', *Political Geography*, 17: 989–1012.

Amy, D. J. (1997) *Proportional Representation: The Case for a Better Election System*. Northampton, MA: Crescent Street Press.

Arrington, T. and Ingalls, G. (1998) 'The limited vote alternative to affirmative districting', *Political Geography*, 7: 701–728.

Baker, G. (1986) 'Judicial determination of political gerrymandering: a "totality of circumstances" approach', *Journal of Law and Politics*, 3: 1–19.

Brischetto, Robert R. and Engstrom, R. L. (1998) 'Cumulative voting and Latino representation: exit surveys in 15 Texas communities', *Social Science Quarterly*, 78: 973–991.

Canon, D. (1999) *Race, Redistricting and Representation: The Unintended Consequences of Black Majority Districts*. Chicago: University of Chicago Press.

Choldin, H. (1994) *Looking for the Last Percent: The Controversy over Census Undercounts*. New Brunswick, NJ: Rutgers University Press.

Cirincione, C., Darling, T. and O'Rourke, T. (2000) 'Assessing South Carolina's 1990s Congressional districting', *Political Geography*, 19: 189–212.

Clark, W. A. V. (1992) 'Local redistricting: the demographic context of boundary drawing', *National Civic Review*, 81: 57–63.

Dixon, R. (1968) *Democratic Representation: Reapportionment in Law and Politics*. New York: Oxford University Press.

Epstein, D. and O'Halloran, S. (1999) 'A social science approach to race, redistricting and representation', *American Political Science Review*, 93: 187–191.

Forest, B. (1995) 'Taming race: the role of space in voting rights litigation', *Urban Geography*, 16: 98–111.

Forest, B. (2001) 'Mapping democracy: racial identity and the quandary of political representation', *Annals of the Association of American Geographers*, 91: 143–166.

Grofman, B. (ed.) (1982) *Representation and Redistricting Issues*. New York: Heath.

Grofman, B. (1990) *Political Gerrymandering and the Courts*. New York: Agathon.

Grofman, B. (ed.) (1998) *Race and Redistricting in the 1990s*. New York: Agathon.

Gudgin, G. and Taylor, P. (1978) *Seats, Votes and the Spatial Organization of Elections*. London: Pion.

Guinier, L. (1995) *The Tyranny of the Majority*. New York: Free Press.

Ingalls, G., Webster, G. and Leib, J. (1997) 'Fifty years of political change in the South: electing women and African-Americans to public office', *Southeastern Geographer*, 37: 140–161.

Johnston, R. and Pattie, C. (2001) 'New labor, new electoral system, new electoral geographies? A review of proposed constitutional changes in the United Kingdom', *Political Geography*, 19: 495–515.

Johnston, R., Pattie, C., Dorling, D. and Rossiter, D. (2001) *From Votes to Seats: The Operation of the UK Electoral System since 1945*. Manchester: Manchester University Press.

Kelly, L. (1996) 'Race and place: geographic and transcendent community in the post-Shaw era', *Vanderbilt Law Review*, 49: 227–308.

Leib, J. (1998) 'Communities of interest and minority districting after *Miller v. Johnson*', *Political Geography*, 17: 683–700.

Lennertz, J. (2000) 'Back in their proper place: racial gerrymandering in Georgia', *Political Geography*, 19: 163–188.

Lublin, D. (1997) *The Paradox of Representation*. Princeton, NJ: Princeton University Press.

Monmonier, M. (2001) *Bushmanders and Bullwinkles*. Chicago: University of Chicago Press.

Morrill, R. (1981) *Political Districting and Geographic Theory*. Washington, DC: Association of American Geographers.

Morrill, R. (1987) 'Redistricting, region and representation', *Political Geography Quarterly*, 6: 241–260.

Morrill, R. (1999) 'Electoral geography and gerrymandering: space and politics', in G. Demko and W. B. Wood (eds), *Reordering the World*. New York: Westview.

Niemi, R. and Deegan, J. (1978) 'A theory of political districting', *American Political Science Review*, 72: 1304–1323.

O'Loughlin, J. (1982) 'The identification and evaluation of racial gerrymandering', *Annals of the Association of American Geographers*, 72: 165–184.

O'Loughlin, J. and Taylor, A. (1982) 'Choices in redistricting and electoral outcomes: the case of Mobile, AL', *Political Geography Quarterly*, 1: 317–339.

Pildes, R. and Niemi, R. (1993) 'Expressive harms, bizarre districts, and voting rights: evaluating election district appearances after *Shaw v. Reno*', *Michigan Law Review*, 92: 483–587.

Pitkin, H. (1967) *The Concept of Representation*. Berkeley: University of California Press.

Raburn, R. and Leib, J. (1994) 'Congressional district building blocks: choice and impact in the 1990s', *Comparative State Politics*, 15: 17–27.

Richie, R. (1998) 'Full representation: the future of proportional election systems', *National Civic Review*, 87: 85–95.

Rossiter, D., Johnston, R. and Pattie, C. (1999) *The Boundary Commissions: Redrawing the UK's Map of Parliamentary Constituencies*. Manchester: University of Manchester Press.

Rush, M. (2000) 'Redistricting and partisan fluidity: do we really know a gerrymander when we see one?', *Political Geography*, 19: 249–260.

Webster, G. (1993) 'Congressional redistricting and African-American representation in the 1990s: an example from Alabama', *Political Geography*, 12: 549–564.

Webster, G. (2000) 'Playing a game with changing rules: geography, politics and redistricting in the 1990s', *Political Geography*, 19: 141–162.

Wildgen, J. and Engstrom, R. (1980) 'Spatial distribution of partisan support and the seats–votes relationship', *Legislative Studies Quarterly*, 5: 423–445.

5 Citizens and the State: Citizenship Formations in Space and Time

Sallie A. Marston and
Katharyne Mitchell

Everyone seems to be talking about citizenship these days. What has happened in recent years to make this concept so relevant in the imaginations of both scholars and the public? In Western democracies, citizenship is often defined as the rights and duties relating to an individual's membership in a political community. The boundaries of this community are most commonly understood as those of the nation-state, and membership implies some degree of integration into a national community and a common heritage. But what happens when that community becomes swept up into global currents that may overwhelm its power and control? What happens when new forms of financial flows, foreign direct investment, transnational migration, supra-national institutions, local autonomy movements, and non-government organisations seem to render the national community obsolete?

Clearly, the model of the discrete, powerful and autonomous nation-state is no longer adequate for describing contemporary geopolitical or geo-economic relations. The more interesting debates now revolve around the kinds of tensions that are erupting as the state's power becomes transformed. A key source of tension remains the concept of citizenship. As many national governments lose the capacity to protect their citizens, at the same time that more and more groups are demanding the expansion and redefinition of citizenship rights, there has emerged an intense and often fractious struggle over the meaning and terms of citizenship worldwide.

One reason that the idea of citizenship is currently under scrutiny relates to broad changes in the ways that many governments have responded to the imperatives of globalisation. In many Western democracies, state policy has moved sharply to the conservative right through a declining belief in the ideology and practice of welfarism. The rights of a nation's residents to certain benefits such as health care, education, shelter and food – rights that were successfully struggled for throughout the last century – are now disputed, and many conservative and middle-road politicians have made broad cuts in entitlements that, until recently, were considered the responsibility of the state. These cuts often encompass the entitlements of both citizens and non-citizens, although non-citizens (legal and undocumented) have been especially targeted. A good example of the latter case can be found in the passage of Proposition 187 in California in 1994. Although eventually overturned by the courts, Proposition 187 initially denied illegal immigrants access to basic schooling, health care and social services, and carried the proviso that civil servants, such as teachers, law enforcement personnel, social security workers or health care staff, were *required* to alert federal authorities if they suspected that one of their clients was not a US citizen or green card holder. A more recent case in the US context is the 1996 Welfare and Responsibility Act, expressly designed to reduce state welfare assistance to the poor. Similar types of bill have been passed in Canada and England in the past decade (see Peck, 2001).

The attack on systems of governance, especially the provision of social entitlements to a nation's residents, is one feature of what has been termed neo-liberalism. Neo-liberalism is an economic and social philosophy of government that is spreading concurrently with other globalisation processes, and is being adopted by an increasing number of nation-states. This 'new' liberalism draws from prior conceptualisations of liberalism in several ways. The incipient liberalism promulgated by British philosophers such as Locke in the seventeenth century was premised on the paramount rights of the individual to procure and own property, to maintain complete control over the body, including the right to contract out labor, and to be protected in these activities by a state which secured, protected and enabled individual economic rights, but which otherwise did not intervene in the processes and practices of everyday life. As the rights of 'free' workers to contract out their labor for a wage was essential for the newly industrialising economy in Britain at that historical moment, many scholars have argued that the philosophical deliberations around the concept of 'liberalism' were intricately interlinked with the economic transformations then affecting British society. Similarly, the formation of the British parliamentary state of the seventeenth and eighteenth centuries, which developed in tandem with liberal philosophies of the rights of the individual, reflected both the rise of capitalism as a socio-economic system, and a strong and abiding undercurrent of distrust for a strong, developmental or interventionist state.

The interventionist or regulatory state was the hallmark of the twentieth century, particularly the period of high Fordism between the Second World War and the late 1970s. As our first case study demonstrates, the ideology and implementation of the interventionist or welfare state was won through difficult and ongoing contestations over social service provisioning, largely at the urban scale. As we will show, however, these victories were uneven, partial and subject to constant pressure, and in the last two decades, many of the social reforms won throughout the twentieth century have begun to be dismantled. Contemporary neo-liberalism, spanning the last two decades is, in many senses, a return to the earlier notion of civic liberalism, especially in its strongly negative reaction to state intervention in any realm aside from the supply-side stimulation of economic production. In addition to the attack on demand-side governance, neo-liberal state formations emphasise the free circulation of capital and commodities and the discrediting of philosophies and practices of state welfarism. It is a philosophy very much in tune with the contemporary form taken by capitalism, a form characterised by increasingly global, unencumbered and flexible systems of production and accumulation.

Most scholars link the concept of citizenship with the rise of liberalism, and also with the shifting formations taken by capitalism over the course of the last three centuries (Turner, 1993). In this chapter we examine the ways in which the state's relationship to citizenship is always shifting, sometimes contradictory and inevitably interrelated with the form and logic of capitalist development. We advocate an approach to citizenship that recognises it, not as a stable and evolving conceptual category, but as a non-static, non-linear social, political, cultural, economic, and legal construction that is best rendered in terms of a *citizenship formation*. The citizenship formation approach recognises citizenship as a process that is both enabling and constraining. Two case studies are described to illustrate this point. The first is an historical geographical case study of turn-of-the-nineteenth-century white, US, urban women's movements. This case is used to illustrate the concept of citizenship formation where popular constructions of citizenship were used to manipulate and reconfigure state power producing the foundations of the modern US welfare state. This nineteenth-century case study shows the state in transition from a *laissez-faire* posture to one where it began to assume increasing responsibility for the social welfare of much of its citizenry. This period represents the emergence of the earliest manifestations of the liberal welfare state. The second is a contemporary case study of turn-of-the-twentieth-century Hong Kong transnational migrants to Vancouver, Canada. These migrants had special access to citizenship claims based upon their high economic status. This case is used to illustrate the impact that transnational migration, and global flows of capital and people in general, is having on contemporary understandings of citizenship. This second case captures the state in transition once again as it moves away from a liberal welfare state to a post-Keynesian or neo-liberal posture, largely abandoning the social welfare responsibilities it had assumed and developed over the course of the previous century. Pairing the two cases enables us to show how the foundations for citizenship are crucially tied to the state and to shifting economic formations, and also how state restructuring and civic resistance or accommodation shifts the terrain of the rights and responsibilities of citizenship.

Conceptualising Citizenship

Citizenship, as it has developed in British and North American societies, owes its legacy to a succession of legal and political rights and responsibilities originating in Britain in the seventeenth century and continuing through to the present. T. H. Marshall, one of the most important early theorists of citizenship, has usefully divided citizenship into three components: civil citizenship, which encompasses civil and legal rights (property rights); political citizenship, which involves the political right to vote, to associate and to participate in government; and social citizenship, which involves social entitlements such as the provisions for health and education (see Marshall and Bottomore, 1992; see also Mann, 1987: 339). Some of the key moments in the early evolution of civil citizenship included the introduction of *habeas corpus*, the Toleration Act, and the abolition of censorship of the press; later they included Catholic Emancipation, the repeal of the Combination Acts, and the resolution to ongoing struggles over freedom of the press (Marshall and Bottomore, 1992: 10). The growth of civil citizenship in the eighteenth century involved fundamental legal changes, as well as changes in cultural attitudes. These shifts included, most importantly, an understanding of the individual's *right to work* – especially the right to form contracts with potential employers. Before this, the right of the individual to contract out his or her labor was not considered a natural right. Individual economic freedom was often won only through a serf's successful escape to one of the rapidly growing 'free' towns: 'The liberty which his predecessors had won by fleeing into the free towns had become his by right. In the towns the terms "freedom" and "citizenship" were interchangeable. When freedom became universal, citizenship grew from a local into a national institution' (Marshall and Bottomore, 1992: 12).

The expansion of individual freedoms into a 'national institution' was one of the key components of the growth of citizenship. It reflected a shift from local, communal relations and social rights rooted in village membership into a sense of a national community and of individual rights guaranteed by a state. This shift in scale from the local to the national and from communally sanctioned rights to those protected by the state is a fundamental aspect of citizenship, and one that is profoundly connected with the rise of industrial capitalism and the rise of modernity in the West. Features of modernity, as scholars such as Turner (1993: vii) have noted, include the decline of particularistic values, the emergence of the idea of a public realm, the erosion of particularistic commitments, secularisation, and the growth of the administrative framework of the nation-state. The push toward modernity and the emergence of citizenship are thus largely equivalent, and both are linked with the growth of individualism and the market economy (see Weber, 1978).

Individualism, as Macpherson (1962: 1) observed, 'has been an outstanding characteristic of the whole subsequent liberal tradition'. It is in the liberal tradition that an understanding of the essence of human freedom as relying on the individual's possession of his or her body and freedom from dependence on the wills of others is derived. The philosophy of liberalism, in the tradition derived from Locke, Hobbes, Bentham and Rousseau, is often touted as the end of 'communitarian' values, because the emphasis on the *contractual* nature of human relations (protected by the state) quickly erodes those of the *communal* (King, 1991: 1). As Turner (1993: 5) noted, 'the emergence of citizenship is in this sense the development of modernity, namely a transition from status to contract'. Early liberal theories, based on what Macpherson describes as a kind of 'possessive' liberalism (the 'ownership' of the body and property), became integral to the concept of citizenship in Western democracies and corresponded precisely with both the growth of a market society and the rise of the bureaucratic modern state. Possessive liberalism was also highly compatible with the advent of the legal/juridical principles of 'civil' citizenship.

If *civil* citizenship was completely compatible with the rise of early industrial capitalism, what was the relationship of a burgeoning market society with the broader rights won under *political* and *social* citizenship in the following centuries? Political citizenship initially involved merely the extension of rights to new sectors of the population. With successive Reform Acts in Britain throughout the nineteenth century, however, it led finally to universal suffrage in 1918. (Suffrage for women was won in the United States in 1921. Many minority groups, such as the Chinese, however, were not fully enfranchised until after the Second World War.) The Act of 1918 allowed political claims to be made on the basis of 'personhood', or a general capacity for citizenship, rather than on purely economic grounds; in other words, it 'shifted the basis of political rights from economic substance to personal status' (Marshall and Bottomore, 1992: 13). Political citizenship thus introduced the concept (if not the full implementation) of political equality, as represented most fully in the right to vote and be represented by a government.

The concept of social citizenship that emerged in the twentieth century followed decades of intense struggle by the working classes to secure fundamental social rights. Social rights, furthermore, were demanded as a fundamental component of citizenship, rather than as an alternative to them. These rights included, most prominently, minimum governmental provisions in the areas of health, education and housing, but extended to encompass a more general egalitarian claim to a level of social reproduction of all individuals above poverty level, and even toward a level of material satisfaction and enjoyment. In Britain the first real advance in social rights began at the end of the nineteenth century, but grew significantly during the years following the Second World War. And in Western Europe and North America, the foundations of a strong liberal welfare state were expanded in those post-war years.

In these societies during the post-war years the essential principle of social citizenship involved an understanding that the state should guarantee a minimum supply of services such as shelter and education and/or a money income necessary for the support of basic living needs, such as an old-age pension or unemployment insurance. For Marshall, however, this was merely the skeletal structure that upheld a far more important element of social citizenship: solving some of the inherent contradictions between capitalism and democracy. Marshall argued that with the advance of social citizenship there would be, to some extent, a retreat of class divisions in society, and that citizenship would form the necessary buffer between the moral and politically active individual and the amoral nature of social exchange in the marketplace. Citizenship could contain the divisive effects of class conflict through its relief of material concerns, but, even more importantly, through its widely shared ideology of the right of all individuals to a share in the social heritage and to live the life of a civilized being according to the standards prevailing in the society (Fraser, 1997; Gordon, 1997: 92; see also Hindess, 1993: 20; Turner, 1993). For Marshall, the ongoing extension of social citizenship provided the entry into a national community of citizens through the provision of a kind of egalitarian 'status' which superseded the type of status heretofore conferred only through class.

There have been numerous critiques of Marshall's conceptualisation of citizenship in the past four decades (see Barbalet, 1988; Hindess, 1993: 23; Kofman, 1995: 125; Mann, 1987; Marston, 1990; Turner, 1986 and 1993: 7; Vogel, 1991: 66). In addition to his general neglect of women and the importance of the state's role in his analysis, the most common source of criticism has focused on the seemingly natural progression made between economic development and the growth of civil society. The argument is often read as teleology of progress, wherein greater political spaces and freedoms are opened up alongside the inevitable overall growth of the economy. For Marshall, the extension of citizenship and political mobilisation were fundamentally connected with national integration. Writing primarily in the 1950s and 1960s, he saw the development of social citizenship as inextricably linked with Britain's nation-building project at the end of the war. The national community and social citizenship were necessarily conjoined, and both would fail if rendered asunder, an unthinkable prospect for a territorially-bounded community as old as Britain's:

> The retreat of social citizenship was thus barely conceivable because it was part of the very foundations of national communities. During the three decades of 'the long boom' – the almost uninterrupted expansion of capitalist economies after the Second World War – that impression of permanence was strengthened because growing prosperity funded an expanding range of entitlements. (Moran, 1991: 33)

However, a more conflict-focused approach to the rise of citizenship suggests that there is no necessary link between economic development and the growth of civil society, or the expansion of political space as a logical outcome of capitalist development. In fact, the frontiers of citizenship can contract as well as expand under different circumstances, including those of war and large-scale migration, as well as economic growth and change: 'Citizenship can be conceived as a series of expanding circles which are pushed forward by the momentum of conflict and struggle. [...] These rights can also be undermined by economic recession, by right-wing political violence, by inflation and by the redefinition of social participation through the law' (Turner, 1986: xii). Despite these numerous limitations, Marshall's theorisation of citizenship is still useful with regard to understanding the shifting connections between capitalism and citizenship and between the state and society. It is not the separation into different types of citizenship or even the periodisation of these types of citizenship that is the most problematic dimension of Marshall's work, but rather the seemingly natural and ongoing progression from a relatively constrained form of citizenship to a more open and democratic one.

Feminist Critiques of Citizenship

The most trenchant and extensive set of contemporary criticism of the concept of citizenship has come from feminist theorists who have pointed to the practical and political problems inherent in the liberal formulation of a gender-neutral, universal agent-citizen. As discussed in the previous section, liberal political theory identified a citizen through a propertied connection to the market. And, for much of the history of Western democracies, the liberal citizenship concept, while embracing the ideal of the equality of individuals and rights, has actually been a political, social, and legal status that has systematically denied access to a whole range of social agents. Because of deep-seated prejudices, such as sexism and racism, predicated on strongly held beliefs about natural roles and predispositions, men with education and property have had access to citizenship, while people of color, and even for a long time poor white men, were denied that same access. In effect, the notion of *popular sovereignty* that is the foundation of the constitutions of many Western democracies is something of a fiction, an exercise in ideological construction about the national community and who really belongs to it (Krasner, 1989). Characterising the notion of popular sovereignty – that is, government by the people, of the people and for the people – as a fiction is meant to expose the yawning gap between the liberal ideal and the actual exclusionary underpinnings of citizenship in practice. At the same time, however, by recognising that in actual practice, access to the rights and responsibilities of citizenship has shifted over times and across spaces, this characterisation admits the possibility, through understanding, struggle and resistance, of closing the gap between the ideal and the real.

The feminist critique of liberalism and the centrally important notion of citizenship is a response to liberalism's connection to Enlightenment arguments about the social contract and the need to found civil society upon reason and rationality (the public life of politics and markets) as opposed to emotion and spirituality (the private life of the home) (see Marston, 1990). Liberalism effectively barred women and others from direct participation in civil society because it was believed that their 'disorderly' nature rendered them unable to be rational and reasonable and to develop a sense of justice (Pateman, 1988). Thus, ideas about the natural roles of men and women have been and continue to be central to the ideological underpinnings of the national community of citizens and their relationship to the state. In pointing out the central failings of the liberal construction of citizenship, feminists have shown how cultural understandings of the natural roles of men and women also directly inform those failings.

The feminist critique of citizenship incorporates a wide range of perspectives. Neo-liberal feminist theories look to the concept of distributive justice as a way of negotiating the naturalised inequality between men and women. And yet, as has been the case for a masculinist liberal discourse, the ethic of justice has not been especially successful in establishing equal rights. The communitarian perspective argues that women should be guaranteed equal access to the public sphere based on their role as caregivers (Elshtain, 1981; Lister, 1997; Ruddick, 1989). But such maternalist thinking romanticises and misrepresents the private sphere as a necessarily female one, essentialising women in the process (Deitz, 1985). The social democratic critique, most thoroughly developed by Nancy Fraser (1997), argues for a wholesale transformation of the gendered socio-spatial structures of everyday life so that men and women can exist as equals in both the public and the private sphere. Each of these approaches, in accordance with their varying political commitments, attempts to redress the shortcomings of the liberal conceptualisation of citizenship. The first two, even as they have attempted to subvert or re-imagine the practices of liberal universalism, have unavoidably perpetuated its problems by failing to come to grips with the many significant and ungeneralisable differences that actually divide women (and men, for that matter) and make the notion of a generic female political subject unworkable, if not entirely unthinkable. The social democratic position, while salutary, requires profound social and political change and is, practically speaking, still a long way from realisation.

While we are deeply committed to the model of transformative recognition (of difference) and redistribution (of social roles) that Fraser advocates, we argue for another way into the paradox of citizenship by starting from a different premise than the ideal, universal subject of liberalism. That is, rather than attempting to restructure liberalism's universal subject with the particularisms of real actors, we argue that we must first understand citizenship as it is actually constructed in specific periods and places, in order to begin to come to terms with the challenges that this potentially transformative political status holds for enabling equality and justice for all. We argue also that one of the most effective ways to comprehend the concept of citizenship is through the interaction between the citizen and the state, and the state's power to define who belongs and who does not belong to the national community. Through our two case studies, we also want to make clear that the relationship between the state and citizens is open to contestation, that conceptualisations of citizenship are open to redefinition through popular struggle. By taking this position we do not mean to imply that the state always yields to pressure from citizens. Rather, we mean to argue that the state responds to powerful social, political, cultural and economic forces such that any citizenship formation is the result of shifting relationships among and between the state, capital, and citizens.

Citizenship Formations

We offer an alternative lens on citizenship, one that emphasizes its deep imbrication within the ongoing processes of state formation and capitalism. The emphasis on citizenship *formation* allows us to think about citizenship as something unfolding and constantly changing, rather than a finished product. Although changing and often contested, its primary definitions generally derive from the state. As may be expected, there is tremendous geographical variety in the ways that different states define and manipulate the concept of citizenship. This variety is also expressed in changes within a single state's definitions over time, especially as different groups claim their 'rights' as citizens or even the right to be considered citizens. The idea of *formation* emphasizes the dynamic and non-linear quality of citizenship, which may expand or contract in different moments depending on the context in which the state is integrated into the global economy, the types of internal battle occurring within state boundaries, or a host of other variables. Thus, in contrast with Marshall's image of an ever-expanding and increasingly inclusive framework of citizenship, the idea of citizenship formation emphasises its elastic qualities, which may indeed expand, but may also snap back or shift shape completely.

One of the central forces affecting citizenship formation is the economy. The state's position in relation to capital is absolutely crucial in understanding how and why citizenship comes to be defined or redefined in the manner it does. This goes for both the local state, urban development and the practices of citizenship at the neighborhood or city scale, as we show with reference to the nineteenth-century case study, as well as to global capital, and the practices of national citizenship, as in our contemporary case study. Indeed, what the two case studies make clear is that the dominant (and often exclusive) emphasis on the scale of the nation-state, so prevalent in political geography, can sometimes actually hide more than it reveals about the contested relationship between citizens and the state. In both of our cases, the national scale is not the key scale in negotiations over political and economic power and the meaning and practice of citizenship. The cases, and our promotion of the concept of citizenship formation, make clear that political and economic restructuring seems consistently to enable new scales for political action and, in this process, allows for a redefinition of the rights and responsibilities of national citizenship.

Indeed, the state's effort to control the meaning of citizenship is always dependent on its relationship with capitalist development and the circulation of commodities and information. It is also, as we shall see, dependent on the shifting relationships between state and nation, state and supra-national forces, and state and local forces. As the state itself is not a monolithic black box, but is composed of innumerable pieces and players, it is also often locked in internal, *intra*-state struggles between different bureaucratic factions operating under the same rubric. Thus, even within the state the concept of citizenship contains elements of contradiction, many of which are worked out as a result of internal struggles taking place over time. In the following two cases we examine citizenship formation in two different historical moments – one liberal, one neo-liberal – hoping to demonstrate its relationship with the forces of capitalism, and its non-linear and dynamic nature. We also emphasise that just as citizenship is dynamic, so too is the state. In this respect, each of the cases effectively highlights a different moment in state formation and restructuring which occurs in response to the cycles of capital accumulation and crisis, as well as to the demands of citizens.

Nineteenth-century Citizenship Formation: 'Female Citizens' and the Welfare State in the United States

Citizenship formation in the late-nineteenth and early-twentieth-century was premised on a widely circulating discourse of domesticity and maternalism and an indirect relationship of individuals to the state. This discourse was a reflection, refraction and response to a whole range of structural transformations in train in urban areas in the United States during this period. At the turn of the nineteenth century in the United States, the state, from the local to the national level, was being restructured as urban pressure groups, largely related to the progressive movement, argued for its increasing role in assuming responsibility for the provision of public services and fostering and protecting social welfare. It should also be recalled that the turn of the nineteenth century was also a period when immigration, industrialisation and urbanisation were increasing dramatically. As a response to these profound changes, and in many ways in an attempt to control and mitigate the perceived negative consequences of them, middle-class women of Western European descent living in cities in the Midwest and the East operated through the woman's movement and its related strands – including voluntary motherhood, domestic feminism, municipal housekeeping, mother's aid and other social provision movements – to reshape the state and create a place for themselves as engaged 'female citizens.'

The relationship to the state that these women possessed was a complicated one. At the turn of the nineteenth century, women (as well as freed slaves, children, and propertyless men) were not considered citizens. Whether married or not, a woman was dependent on either her father or her husband to represent her to the outside world in political and legal issues. In the case of a wife, her identity stemmed from her husband's so that through marriage, she and her husband constituted one person, the husband, who controlled both her property and her legal identity. In short, for women of this period, membership in the republic was derivative, disqualifying them from participating in the political life of voting, sitting on juries, or taking legal action in their own names. In fact, bourgeois liberalism, which was at the foundation of republican ideals, delegated different social roles to men and women based on their 'essential natures'. Women's role was confined to the private sphere of the home and men's to the public life of politics and markets. Following over a hundred years of republican government and a liberal ideology of separate gendered spheres, Victorian urban middle-class women began to understand their responsibilities for caring for the family and home as an ontological and ideological position for approaching, albeit unconventionally, the male-dominated public sphere of local and national politics.

Using their identities as mothers and housewives, and arguing that the city was an extension of the home, these women contended that their moral superiority as mothers (based on their exclusion from the public sphere) and the expert knowledge they had developed as household managers, enabled them to uniquely comprehend and offer reasonable solutions to the problems that attended the increasing urbanisation of capital. The discourse that surrounded women's activism emerged from an understanding of their role as mothers and their responsibilities to rear and cultivate their children as citizens of a new democracy. They called themselves 'female citizens', a hybrid category that acknowledged their political, social, and civil distance from the official category of citizenship at the same time that it made a new political space for them to voice their insights and demands. What is perhaps most interesting about their hybrid status is that although turn-of-the-nineteenth-century American women were not citizens in law or practice, and therefore did not enjoy the particular rights and entitlements of citizenship, they were effective in arguing for their political right to deploy a female understanding of the home to address wider social concerns and to define and pursue political responsibilities through a category that was widely used and widely accepted as legitimate.

The means through which the 'female citizens' of the urban middle class pursued the self-assumed rights and responsibilities that enabled them to have a significant influence over state functions were many, but the most well documented are the various groups of the period that were related directly and indirectly to the progressive movement. The progressive movement involved shifting coalitions and loosely organised groups operating on any number of issues and concerns. It encompassed both men and women and, as a largely urban movement, was aimed at reforming and utilising the state as a provider and guarantor of justice and social welfare. The progressive movement, and closely and distantly related strands, was devoted to enhancing the power of the state to produce new sets of rules and practices to govern new populations and address new problems. The domestic feminism movement, for instance, was very much predicated on extending the rules that 'governed' middle-class households – rules about personal hygiene, food preparation and eating, marriage, childrearing, and health care – beyond the bounds of the private domicile into the city, pushing local government to become increasingly involved in creating laws and enacting practices that would rationalise the city in many of the same ways that the middle class had rationalised the space of the middle-class home. Much of the rhetoric of efficiency, standardisation, and sanitation that characterised the popular domestic feminism texts of the period, widely consumed by middle-class urban women, was repeated and extended to apply to the burgeoning American city, seen to be teeming with new immigrant populations, noxious industries, and corrupt businesses and political machines.

What in the late nineteenth century had been locally organised and focused organisations, such as the domestic feminism movement, became in the early twentieth century, coordinated, nationally organised and directed organisations predicated on middle-class expertise with its impulse for standardisation and professionalisation. Female urban reformers were pivotal in the progressive movement and they wielded a great deal of political and social power through organisations such as the National Associated Clubs of Domestic Science, the National Civic Federation, the National Child Labor Committee, the National Municipal League, and the National Playground Association of America (Chudacoff and Smith, 2000). Thus, by comprehending their identity as female citizens, albeit ones without the formal legal, social and political rights guaranteed to white property-owning men, urban middle-class women constructed for themselves a powerful relationship to the local and the federal state that enabled them to operate as citizens without actually being ones.

In addition to addressing issues of social welfare and service provisions at the local level, urban middle-class women by the early twentieth century were also able to influence, quite substantially, the restructuring of the federal state to assume increasing responsibility for the social well-being of its citizens, especially women and children. Through organisations such as the National Child Labor Committee and the US Women's Bureau, women's groups were able to promote the creation of modern public welfare by spreading the notion of Mother's Aid from state to state so that by the end of the Depression, the federal government nationalised it by creating Aid to Dependent Children (ADC). It is widely accepted among scholars of gender that the modern US welfare state originated in 1915 with these state-level mother's pension programs (Gordon, 1997).

As Gwendolyn Mink (1995) has shown, the proponents of Mother's Aid drew from an evolving tradition of activism that moved from 'the republican ideology of gender citizenship's to the separate spheres ideology of the Victorian period to the mothers' politics of the late nineteenth and early twentieth centuries which asserted women's political significance to "the race" and "the nation". By understanding themselves as experts of the private sphere with knowledge valuable to improved functioning of the public sphere, social and policy activists connected the health of the polity to the quality of motherhood and demanded that government provide economic assistance to poor women and children. Molly Ladd-Taylor summarises maternalism as an ideology and a discourse whose adherents hold:

(1) that there is a uniquely feminine value system based on care and nurturance; (2) that mothers perform a service to the state by raising citizen-workers; (3) that women are united across class, race, and nation by their common capacity for motherhood and therefore share a responsibility for all the world's children; and (4) that ideally men should earn a family wage to support their 'dependent' wives and children at home. (Ladd-Taylor, 1994: 3)

It is important to understand that the Mother's Aid policies were not simply altruistic gestures but were very much predicated, not insignificantly, on class and race-based interests. The rapid industrialisation of Northern and Midwestern cities, as well as the growing numbers of foreign immigrants and freed slaves who poured into them seeking livelihoods, created serious anxieties among middle-class men and women who perceived a threat to their way of life. The social welfare policies that resulted from the activism of female citizens were in no small part attempts at the social control of the poor. Debates raged about whether to restrict immigration and segregate former slaves and new immigrants or to attempt to incorporate the new groups into white middle-class norms institutionalised through policies like Mother's Aid (Gordon, 1997). Women social and policy activists worked hard to achieve the latter, an approach that dominated American social welfare policy for nearly an entire century. Turn-of-the-century white middle-class urban women constituted a powerful citizenship formation that wielded substantial political power over the state without actually possessing the formal identity of citizens. Indeed, it is widely understood that their intense activism ultimately paved the way, as the twentieth century wore on, for women, both native-born Americans and immigrants (as well as African Americans), to gain access to the social, political, and legal rights of citizenship that had been available to white property-owning men for 150 years.

This case provides an illustration of citizenship empowerment, without the actual rights attendant to formal citizenship. At the same time, however, that new rights and privileges were enabled by the transformation of state rules and the construction of new state bureaucracies, the opening up of citizenship for white, urban women closed off access to this evolving citizenship formation to other, poor, non-white, immigrant urban and rural populations. It also demonstrates some of the ways in which new rules and regulations governing women and children developed with respect to the state as it assumed increasing responsibilities for the welfare of these new social groups. The case study that follows provides both an endorsement of and a contrast to this one. Both cases illustrate how powerful social actors were able to use their class positions to reshape and redirect state functions to embrace and extend benefits to new constituencies. Both also reveal how the activism that results in new state practices and understandings of citizenship can emerge from scales above and below the national one. Both cases also refer to significant periods of state, capital and citizenship restructuring, periods of substantial immigration and demographic transformation.

The two case studies reveal important contrasts as well. The first one exposes the emergence of the liberal (welfare) state and the ways in which it became enjoined by powerful middle-class actors to create new bureaucracies to support the social reproduction of its citizenry. The second case study discloses the emergence of the neo-liberal (entrepreneurial) state and the ways in which it has been pressured by a different set of powerful transnational actors to open up new opportunities in support of capitalist production (relegating its previous commitments to social reproduction to the marketplace). And both cases focus on different bureaucracies within the state apparatus: collective consumption and welfare in the first one, and economic development and immigration in the second.

Twentieth-century Citizenship Formation: Hong Kong Business Immigrants and the Canadian State

In addition to the cross-border economic flows characteristic of globalisation, flows of people across national borders have put major pressure on both the policy and the philosophy of citizenship. Over the past four decades, mass migration has been unprecedented in scope and intensity (Castles and Miller, 1998), leading to the formation of multi-ethnic societies around the globe, and also to increasing demand for and resistance to enhanced citizenship rights and duties. Migration, furthermore, is no longer limited to the two most common models of circular migration or assimilation. In recent years a major phenomenon termed 'transnational migration' represents a new model, a pattern of migration wherein migrants live bi-nationally, setting up homes and places of work in more than one nation-state (Guarnizo, 1994; Rouse, 1991).

Transnational migration has enormous implications for the concept and practice of citizenship, engendering a new way of thinking about national identity and belonging among the migrants themselves, as well as producing new sets of issues *vis-à-vis* state policy. For example, for some of the poorer states, such as Mexico, the Dominican Republic and India, the economic remittances that transnational migrants send back 'home' have become an indispensable part of the receiving country's economy. Rather than risk losing the allegiance of these migrants, many of whom have established jobs and residences in the cities of core countries, the poorer states will offer benefits such as 'dual citizenship', which, in addition to conferring an ongoing sense of national identity and commitment, can also confer important material benefits, such as the right to inherit property.

Dual citizenship is just one example of the many reforms that are being proposed or implemented as a result of the new tensions characteristic of global restructuring. The state, in a sense, has become deterritorialised, such that its foundation upon a national community no longer corresponds directly to its territorial borders. Dual citizenship confers identity within a national community without the necessity of residing or working in that community, and with the legal agreement national identity might be shared with another nation. In terms of state policy, this is a fundamental shift from the past, where nearly every nation demanded strict, undivided allegiance.

In this second case study we look at the movement of tens of thousands of people in the 1980s and 1990s, who left Hong Kong for Canada primarily as a result of Hong Kong's pending transfer of control from Britain to the People's Republic of China in 1997. In Canada, in addition to the 'normal' immigration process based on a system of points, a special Business Immigration Program was added in the 1980s that allowed wealthy investors and entrepreneurs to receive a visa and enter the country ahead of the general processing queue. (The average amount of waiting time for 'regular' applicants from Hong Kong was several years). The Business Program, established in 1978, but greatly expanded in 1984 and 1986, was explicitly designed to attract those who could contribute to Canada's economic development. As of 1991, investors in this program were required to have a personal net worth of at least C$500,000 and promise to invest C$350,000 in a Canadian venture. Entrepreneurs were expected to have a track record in business and to open up a business in Canada employing at least one Canadian (Nash, 1992). After three years of residence in Canada, the immigrants could apply for Canadian citizenship.

For many years after the program's establishment, Hong Kong led as the primary source country for business immigrants to Canada. For the city of Vancouver, British Columbia, the number of people arriving from Hong Kong averaged around 10,000 for the years between 1990 and 1996, with the largest number, 15,663, entering in 1994. Of these, approximately one-quarter entered through the Business Immigration Program. The amount of capital brought in by this group was astonishing, with overall capital flows into the city of Vancouver estimated at over C$1 billion per year during this time period (Mitchell, 1993). Owing to their wealth and status, and to the capital influx accompanying them, this immigrant group rapidly became quite prominent, and in many cases their arrival also caused tremendous conflict in a number of economic, social and cultural arenas.

One of the most serious sources of tension in Vancouver was the destruction of trees and natural landscaping, and the construction of extremely large houses on relatively small lots. These changes, which were linked in residents' minds with the arrival of the wealthy Chinese immigrants, occurred in formerly white, upper-middle-class neighborhoods. The rapid transformation of these neighborhoods (including major increases in apartment rates and house prices) provoked tense discussions about immigration, citizenship, and what it meant to be Canadian. For the first time, many long-term Vancouver residents and Canadians questioned state policy and philosophy around the issue of citizenship, something that was largely taken for granted up until this point. For the first time, the state's legitimacy in defining immigration and citizenship policy was interrogated and found wanting.

The very visibility of the business immigrants produced tremendous conflict over the official state discourse on the rights and obligations of citizenship. On the one hand, the economically attractive course of bringing in wealthy immigrants pointed to a greater 'open' border discourse, where early or easily attained citizenship rights could be offered as an incentive for this desirable group. On the other hand, the socio-political agenda of the conservative Mulroney administration in Canada also pointed to a 'closed' border discourse, where citizenship rights should be denied, owing to the so-called 'tide' of migrants who might sap the resources of the state. Thus, in many different instances, the state was caught issuing contradictory statements and policies about immigration and the meaning of citizenship in Canada.

A second point of difficulty for the state arose as a result of the obvious collusion between state policy and the interests of capital. The very introduction of the 'business' category of migration reduced the legitimacy of the state as an autonomous entity. Through the 'selling of passports' to the highest bidder, citizenship was rendered both implicitly and explicitly a commercial venture, rather than a political status. The cozy relationship between the market and the state thus rendered suspect the independent status claimed by the state. In addition, the narrative of progress *vis-à-vis* the ongoing extension and inclusion of citizenship through time was disrupted by the selective rights offered prospective citizens who were wealthy, and the lower status given to the poor. With the introduction of the Business Immigration Program, the message was of a return to citizenship primarily for property owners. Just as in the Greek city-state model and the era of civil liberalism, citizenship was limited to the privileged classes who, on the basis of property ownership, were granted the rights to participate in the governance of the society. With the erosion of citizenship rights for the many, and the extension of certain kinds of rights for the few, the national narrative of an inevitable linear progress in citizenship was lost. Furthermore, this loss was clearly tied to the extension of market forces into every aspect of state activity.

There were also clear contradictions not just within a broader state agenda relating to the meaning and implementation of citizenship, but also between various state sectors. In the era of 'fast' capitalism, with a constant movement of bodies and capital across state borders, it became clear that the state was not interested in guaranteeing the general rights of citizenship to all those within its territorial borders. Citizenship protection, in terms of its entitlements and obligations, became increasingly uneven, selective and fragmented, not related to territory *per se*, but more explicitly to economic considerations. The state thus extended citizenship rights to some but not others, that is, to those who could bring various kinds of advantage to particular state sectors either economically or in terms of state legitimacy. State protection waxed and waned depending on historical and geographical context, and citizenship itself thus began to be perceived as a strategic category that was neither universal nor timeless, but rather one that was easily and often manipulated.

Finally, in terms of political hegemony, state control over the definitions and categorisation of citizenship declined with the admission of a wealthy, high-status group who were increasingly able to create their own definitions and meanings. The wealth of many Hong Kong transmigrants facilitated their ability to counter state discourses and contest the normative assumptions of citizenship, especially those tainted by racist stereotyping, which were promulgated by the state. Increasingly, they repositioned *themselves* in new ways. New kinds of self-referencing by wealthy Hong Kong immigrants as bridge-builders, model minorities, Pacific Rim citizens and economic subjects, all indicated the manner in which immigrants were *actors* and not merely receivers of state-driven citizenship categories and norms.

Indeed, many individuals began to be selective as to what aspects of citizenship they wished to uphold. Civil citizenship, for example, emphasising the rights of the free, unencumbered individual rather than social or political rights, was often well suited to the transnational lifestyle of *homo economicus* immigrants (Ong, 1999). In one particular struggle over a downzoning issue in a Vancouver neighborhood, for example, recent Hong Kong immigrants argued for their rights as property owners, invoking the language of civil liberalism to attack local state regulations of house style and size. One Chinese-Canadian speaker said at the public hearings in 1992:

> I live in Shaughnessy and we built a house very much to my liking. The new zoning would not allow enough space for me. [...] I strongly oppose this new proposal. Why do I have to be inconvenienced by so many regulations? This infringes my freedom. Canada is a democratic country and democracy should be returned to the people. (cited in Ley, 1995)

This example demonstrates the manner in which the entry of a particular group of migrants in a specific time period was a galvanising force in reworking citizenship formation in Canada. The case provides an example of the processual and contested nature of citizenship, the interconnections of citizenship formation with economic processes, and the types of inevitable and ongoing contradiction inherent *within* the state. It also demonstrates some of the contradictions between neo-liberal state practices of unfettered movement (especially for capital), and national narratives of coherence and stability. These contradictions, moreover, were both exacerbated and rendered more highly visible as a result of the process of transnational migration itself, as well as the practices of the Hong Kong immigrants after their arrival in Canada.

In the early history of citizenship formation in the West, freedom of trade and freedom of the individual were considered to be bound together in creating prosperity and happiness for the community (Marshall and Bottomore, 1992: 11). This community was conceived as the *national* community at a key moment of nation building. As political rights became established, this sense of a national community, territorially bounded by the nation-state, remained. Political citizenship, and then social citizenship, was assumed to be contained within a territorially defined national space. With the arrival of new kinds of politically savvy, transnational actors such as the Hong Kong Chinese immigrants, however, this normative spatial connection was increasingly contested, and the rights and duties of citizenship in Canada are now no longer so tightly tied to the idea of a community bounded by state borders.

This twentieth-century case study describes how a citizenship formation has moved from a locally-based sense of community to a national scale, and then on to the transnational. But, ironically, the transnational community in many ways looks like the old local connections – as it is often based on communal relations and sanctions rather than state regulatory institutions. Community is thus defined by things other than literal spatial contiguity. In some cases it may be defined by ethnicity, hometown origins, or language. The loss of hegemonic understandings of a national community as being necessary to political participation, and hence democracy, is just one of many ramifications of this contemporary citizenship formation.

Conclusion

We believe that the notion of citizenship formation helps us to come to terms with the complex ways in which the category of citizenship has developed and changed since it was first expressed as a key Enlightenment concept. Indeed, our cases make clear that citizenship is a category that is often under construction, particularly during periods of large-scale structural transformation. That a category of such central political, economic, social and legal importance is so flexible is noteworthy. More important, though, is an understanding of how those changes have occurred, and how further changes might be possible, especially transformations that enable greater equality and justice for diverse and dynamic populations. Toward that end, we offer several basic observations, derived from our case studies, about the processes and structures involved in the social production of citizenship formations.

The first observation is that citizenship formations are a product of the interaction of civil society *and* the state. Citizenship is not a monolithic social category that is determined by state edict and endures unchanged through time and across place. It is, instead, an actively created and negotiated status that is shifted and remodeled in response to large and small economic, social and cultural processes and movements. Both of our case studies show how the interaction of the state and social actors can produce new understandings of citizenship that incorporate new groups, sometimes in more expanded ways, and in other times in less open ones. For instance, the nineteenth-century case study demonstrates that urban middle-class women were social actors with limited access to the formal rights and responsibilities of citizenship. Yet they were able to influence the shape and direction of state practices and policies and in the process create new spaces for themselves and others as citizens.

The second observation is that geographical scale is centrally implicated in producing and sustaining citizenship formations. Civil society and the state can interact at all levels of social life – from the local to the global – such that the discourses and practices of citizenship can derive from one scale and be effective at another. Which scales are central to the production of a particular citizenship formation is dependent upon the particularities of different historical-geographical moments. Our late-twentieth-century case study shows how the transnational scale and the local scale were coeval in the production of that citizenship formation. In that case, powerful transnational economic actors were able to interact and exhort the national government to make a new space for them as Canadian citizens as well as active participants in the social, political, and economic life of the city of Vancouver.

The third observation is that socio-spatial organisation profoundly affects and is affected by the discourse and practices of citizenship formations. The transforming political economies of the United States and Canada during these different but somewhat parallel historical periods were articulated through new social and spatial arrangements expressed in the expansion and transformation of urban form and the spatial clustering of populations. In both cases, the arrival of new immigrants, the availability of new technologies, and the restructuring of cities and regions further contributed to the production of new ideas and discourses about the national (or transnational) community. New citizenship formations were a byproduct of these socio-spatial changes.

We want to end this chapter on a cautionary note. We want to make it clear that we in no way understand the emergence of new citizenship formations as always improving upon or expanding the possibilities for greater equality and justice for all. Our intent is to be far more conservative, taking the position that new citizenship formations simply signal that significant openings in the social constructions of citizenship can and do occur. It is incumbent upon scholars of citizenship to probe further the notion of citizenship formations in order to delimit the conditions under which greater equality and justice for all might be possible.

References

Barbalet, J. M. (1988) *Citizenship: Rights, Struggle and Class Inequality*. Minneapolis: University of Minnesota Press.

Castles, S. and Miller, M. (1998) *The Age of Migration: International Population Movements in the Modern World*. New York: Guilford Press.

Chudacoff, H. and Smith, J. (2000) *The Evolution of American Urban Society (Fifth Edition)*. Englewood Cliffs, NJ: Prentice Hall.

Dietz, M. G. (1985) 'Citizenship with a feminist face: the problem with maternal thinking', *Political Theory*, 13: 19–37.

Elshtain, J. B. (1981) *Public Man, Private Woman: Women in Social and Political Thought*. Princeton, NJ: Princeton University Press.

Fraser, N. (1997) *Justice Interruptus: Critical Reflections on the 'Postsocialist' Condition*. New York & London: Routledge.

Gordon, L. (1997) *Pitied but not Entitled: Single Mothers and the History of Welfare, 1890–1935*. New York: The Free Press.

Guarnizo, L. (1994) '*Los Dominicanyorks*: The Making of a Binational Society', *Annals of the American Academy of Political and Social Science*, 533: 70–86.

Hindess, B. (1993) 'Citizenship in the modern West', in Turner, B. (ed.), *Citizenship and Social Theory*. London: Sage.

King, D. (1991) 'Citizenship as obligation in the United States: Title II of the Family Support Act of 1988', in Vogel, U. and Moran, M. (eds), *The Frontiers of Citizenship*. New York: St Martin's Press, pp. 1–31.

Kofman, E. (1995) 'Citizenship for some but not for others: spaces of citizenship in contemporary Europe', *Political Geography*, 14 (2): 121–137.

Krasner, S. (1989) 'Sovereignty: an institutional perspective', in Caparoso, J. (ed.), *The Elusive State: International and Comparative Perspectives*. Beverly Hills, CA: Sage.

Ladd-Taylor, M. (1994) *Mother–Work: Women, Child Welfare, and the State, 1890–1930*. Urbana & Chicago: University of Illinois Press.

Ley, D. (1995) 'Between Europe and Asia: the case of the missing sequoias', *Ecumene*, 2: 187–212.

Lister, R. (1997) 'Citizenship: toward a feminist synthesis', *Feminist Review*, 57: 28–48.

Macpherson, C. B. (1962) *The Political Theory of Possessive Individualism: Hobbes to Locke*. Oxford: Clarendon Press.

Mann, M. (1987) 'Ruling class strategies and citizenship', *Sociology*, 21 (3): 339–354.

Marshall, T. H. and Bottomore, T. (1992) *Citizenship and Social Class*. London: Pluto Press.

Marston, S. A. (1990) 'Who are "the People"?: Gender, citizenship, and the making of the American nation', *Environment and Planning D: Society and Space*, 8: 449–458.

Mink, G. (1995) *The Wages of Motherhood: Inequality in the Welfare State, 1917–1942*. Ithaca, NY & London: Cornell University Press.

Mitchell, K. (1993) 'Multiculturalism, or the united colors of capitalism?' *Antipode*, 25 (4): 263–294.

Moran, M. (1991) 'The frontiers of social citizenship: the case of health care entitlements', in Vogel, U. and Moran, M. (eds), *The Frontiers of Citizenship*. New York: St Martin's Press, pp. 32–57.

Nash, A. (1992) 'The emigration of business people and professionals from Hong Kong', *Canada and Hong Kong Update*, Winter: 2–4.

Ong, A. (1999) *Flexible Citizenship: The Cultural Logics of Transnationality*. Durham, NC: Duke University Press.

Pateman, C. (1988) *The Sexual Contract*. Stanford, CA: Stanford University Press.

Peck, J. (2001) *Workfare States*. New York: Guilford Press.

Rouse, R. (1991) 'Mexican migration and the social space of postmodernism', *Diaspora*, Spring: 8–23.

Ruddick, S. (1989) *Maternal Thinking: Towards a Politics of Peace*. London: Women's Press.

Turner, B. (1986) *Citizenship and Capitalism*. London: Allen and Unwin.

Turner, B. (1993) 'Contemporary problems in the theory of citizenship', in Turner, B. (ed.), *Citizenship and Social Theory*. London: Sage.

Vogel, U. (1991) 'Is citizenship gender-specific'?, in Vogel, U. and Moran, M. (eds), *The Frontiers of Citizenship*. New York: St Martin's Press, pp. 58–85.

Weber, M. (1978) *Economy and Society*. Berkeley: University of California Press.

6 Open Borders and Free Population Movement: A Challenge for Liberalism

David M. Smith

At some future point in world civilization, it may well be discovered that the right to free and open movement of people on the surface of the earth is funda- mental to the structure of human opportunity and is therefore basic in the same sense as free religion, speech, and the franchise. Such a conclusion will surprise some, coming at a time when territorial boundaries are guarded more tightly than ever against individual penetration. (Nett, 1971: 218)

More than thirty years ago, Roger Nett published a paper, the title of which described the free movement of people on the face of the earth as 'the civil right we are not ready for'. Departing from the legalistic, conventional and naturalis- tic conceptions of rights which prevailed at the time, he saw rights as *discovered*, not some fixed category but defined and redefined in human social contexts, as part of socially constructed reality. Thus, 'commensurate with the times, we find ourselves discovering new rights, meaning new fundamental conditions which insure some amount of justice and equity among people' (Nett, 1971: 216). For example, we discovered some time ago that free speech was necessary to prevent a monopoly of ideas and a negation of minority opinion. Similarly, the right to vote was discovered to be essential for widening opportunities. Formal expressions of human rights, like the European Convention and post-apartheid South Africa's Bill of Rights, represent recent steps in this ongoing process of attempting to give greater substance to rights, as a fundamental feature of liberal democracy.

Nett suggested that people have always recognized a *de facto* right of migration, to avoid persecution and seek better opportunities:

At the most general level, the understanding is that people do have a right to move away from oppression and also have the right to move away from poverty if they can manage to do so. It takes but little extension of logic to suppose that people ought to have the right to move freely over the surface of the earth as long as they are willing to obey local laws and respect local customs. [...] The mute subjects of the unlimited monarchies of the past and the physically detained subjects of our time, whether or not kept mute, are alike in being arbitrarily locked into a pattern of existence against their will or better interests, so that they feel justified in breaking the pattern. (Nett, 1971: 219)

While movement may involve any spatial scale, the issue is usually posed with respect to migration across international boundaries, from one sovereign state to another. Nett (1971: 223) admitted that advantaged people at present may not be ready to share anything like equal opportunities with the disadvantaged, hence the failure to recognize a formal, *de jure* right to free movement. But he warned that the cost of continuing a world in which so many people are discontented could be greater than that of extending opportunities. Thus, on prudential grounds alone, the time for the discovery of a basic human right to free movement may not be far off.

It is worth considering how things have, or have not, changed in the three decades since Nett originally raised this issue. He recognized a moment in history when migration was more restricted than ever, but in order to keep people in rather than out. Then, a major concern of Western democracies was the constraints imposed on people seeking to escape political oppression in the communist countries of the Soviet Union and its Eastern European satellites. Thirty years on, the main preoccupation of Western Europe has become the possibility of being inundated by refugees from the poverty and insecurity of post-socialism to the East. As Ann Dummett (1992: 180) pointedly observes with respect to the border between Austria and Hungary, after the Hungarian authorities destroyed the wire fence on their side following the end of the communist regime, the Austrian authorities introduced new guards and barriers on their side to stop immigration. During the 1990s ethnic conflicts which lay dormant under communism erupted with brutal ferocity in the Balkans, to generate large flows of people trying to escape conditions far more frightening than those endured by most of them during the socialist era. Elsewhere, the Australia that welcomed 'Boat People' from Vietnam in the 1970s, saw a Liberal Party standing for re-election in 2001 on an anti-immigration platform. The same happened in Denmark.

At the time Nett wrote, the asylum seeker attracted honorific status in countries like Britain, eliciting images of political dissidents courageously challenging some repressive regime. A distinction was made with the economic migrant, whose motive was taken to be less elevated, and whose entry tended to be discouraged. But in recent years the difference has become blurred, with the terms 'asylum seeker' and 'refugee', often prefaced by 'bogus', taking on a pejorative tone. Reflecting something akin to moral panic, one newspaper, identifying the 'hardest-hit' part of the country, refers to the need to 'spread the burden', as refugees 'pour into' a Britain 'swamped' with applications: 'soft-touch' Britain could soon be the world's top destination for people 'most of whom aren't real refugees' (*News of the World*, January 21, 2001). People claiming political asylum may be detained, stigmatized, subject to verbal and physical abuse, and treated like criminals. The heightened fear of terrorism provoked by the attack on the World Trade Center in New York in September 2001, and its aftermath, has greatly increased suspicion of different others in our midst. The growing numbers seeking refuge in Britain have provoked political as well as popular reactions, posing fundamental questions concerning the meaning of national identity and citizenship.

British attitudes to immigrants are, now, almost entirely negative. At the time when Nett wrote, dissident escapees from communism were frequently fêted, with their prominence in the arts and sciences considered assets to national life. Largely overlooked today is the contribution of an earlier generation of immigrants, from the West Indies and the Indian sub-continent, who filled menial jobs shunned by British workers. And it goes almost without mention that British creative and intellectual life would have been much the poorer but for pre-war immigrants fleeing Nazi Germany. For example, the modern movement in architecture was significantly advanced by dozens of immigrants, most of them Jewish, whose individual biographies incorporated various combinations of the asylum seeker and economic migrant (Benton, 1995).

Important changes have also occurred in the wider context within which population migration takes place. Attention has been drawn to the growing disparity between the mobility of population and capital in this era of globalizing economic relations. A definitive collection of papers on the subject introduces the issue with the following reminder: 'Every state claims the right to determine who shall be permitted to enter its territory and almost all exercise this right vigorously [yet] most countries place few restrictions on the transfer of capital into them' (Barry and Goodin, 1992: 3). There is much talk of the demise of the nation-state in the face of growing internal divisions in such forms as ethnic revivals, regional separatism and devolution, and of the external delegation of state powers with the rise of supra-state and international organizations. Boundaries are portrayed as increasingly porous, with postmodern thinking privileging fluidity and movement over fixed spatial forms in the emerging age of 'post-sovereignty' (Shapiro, 1994). Some see citizenship becoming multiple and multi-layered, with different and only partially overlapping personal allegiances, rights and responsibilities.

Another aspect of change over the past three decades has been the ethical basis of debates in political philosophy. In the same year as Nett's paper, John Rawls published his liberal-egalitarian theory of justice (1971), which has so greatly influenced Western political thought ever since. Nett anticipated a growing interest in issues of border control and population movement, hitherto largely neglected by philosophers, which was to take on renewed significance in the last decade of the twentieth century in the context of debates on inclusion, exclusion and citizenship. While there was a large prudential element to Nett's argument for a right to free movement, with its recognition of the cost of not extending opportunities to the disadvantaged, liberal egalitarianism has subsequently provided a philosophical rationale robust enough to withstand much of the critique from alternative paradigms. The fact that the actual practice of supposedly liberal states has not caught up points to the prescience of Nett's early social-constructive account of rights, now commonly recognized as 'not transcendental universal norms for behaviour handed down from on high but human discoveries about our human condition made in the course of human conflict' (Low and Gleeson, 1999: 30). However, lest the notion of discoveries might imply revelation of part of the fabric of the world in some linear path of ethical progress, to describe rights as social *creations* would better reflect contemporary understanding.

The preoccupation of this chapter is with the moral-philosophical case for open borders and free movement, as part of geography's recent engagement with ethics (Proctor and Smith, 1999; Smith, 2000b). It is important to stress at the outset that the issues raised here are crucial to both the theory and practice of liberal democracy. Exercising the power to determine who belongs to and may participate in a political community, and who is permitted (or otherwise) to join them, is a matter of contemporary experience played out in the development and implementation of public policy, and subject to the moralizing discourse of sustained media scrutiny. The ongoing resolution of this process has profound implications for our understanding of the changing spatiality of what we suppose to be democratic politics.

Liberal Egalitarianism and the Place of Good Fortune

[M]any of the societies which seem most fully committed to liberal egalitarian ideals (the welfare states of Western and Northern Europe, including particularly the Scandinavian countries) have quite restrictive immigration and citizenship policies. [...] Unlike other widely acknowledged basic human rights (such as rights protecting free speech, or freedom of religion or rights protecting against arbitrary arrest and imprisonment), claims of an unrestricted right of free movement across borders are rarely, if ever, found in standard enumerations of rights in the constitutions or legal systems of liberal egalitarian countries. (Woodward, 1992: 63)

[N]othing could be more suspect, from the standpoint of social freedom, than disadvantages maintained by coercively regulated borders. (Miller, 1992: 300)

The liberal egalitarian case for free population movement has its origin in the proposition that most, if not all, sources of inequality among persons are contingent and therefore morally arbitrary. Individual advantage is very much dependent on chance, or what I refer to in an earlier account as the place of good fortune (Smith, 2000a; see also Smith, 2000b: 139–42). The point of departure is John Rawls's theory of justice.

Rawls began with the conventional system of natural liberties, in which careers are open to the talented, with all persons having equal opportunities in the formal sense of the same legal rights. However, there is no attempt to promote equality in background social conditions. The initial distribution of income and wealth, for example, is the cumulative effect of prior distributions of natural talents and abilities, as developed or otherwise by social circumstances and such contingencies as accident and good fortune (Rawls, 1971: 72). The injustice of such a system is that it permits access to positions of advantage and distributive shares to be influenced by factors that are arbitrary from a moral point of view, in the sense that they cannot ground claims based on desert. He therefore invoked the principle of 'fair equality of opportunity', under which persons with the same talent and ability and the same willingness to use them should have the same prospects regardless of their initial place in the social system. However, this conception also appears defective, for even if it eliminates the influence of social contingencies, it still permits the distribution of wealth and income to be determined by the natural distribution of abilities and talents, which is arbitrary from a moral perspective: 'There is no more reason to permit the distribution of income and wealth to be settled by the distribution of natural assets than by historical and social fortune' (Rawls, 1971: 73–4).

Erasing the distinction between social-environmental circumstances and natural attributes achieves 'democratic equality', which strongly suggests equality of outcomes. In effect, Rawls made all sources of differential occupational achievement morally arbitrary. There is no case at the most basic level of justification for anything except equality in the distribution of Rawls's primary goods of liberty and opportunity, income and wealth, and the bases of self-respect. Brian Barry (1989: 226) explains the situation in terms of lotteries: the natural lottery which distributes genetic endowments; the social lottery which distributes more or less favourable home and school environments; and the lottery that distributes illness, accidents, and 'the chance of being in the right place at the right time'.

What became known as the argument from arbitrariness, or luck egalitarianism, featured prominently in subsequent work on social justice. And, as Barry implies, to the contingencies explicitly recognized by Rawls can also be added the contingency of geographical place. John Baker explains:

> So much of what people achieve is a matter of being in the right place at the right time, of having good luck in family, teachers, friends, and circumstances, that no one is in a strong position to take *much* credit for the way their lives turn out. There is no such thing as a literally self-made man [*sic*]. And so any judgement of desert will have to look closely at where responsibility really lies. (Baker, 1987: 60)

The reference to responsibility reflects critiques of the argument from arbitrariness that stress the role of individual choice and effort, or what people are able to make of their lives whatever their circumstances. Those unwilling to assign all personal attributes to morally arbitrary fortune are concerned to leave something to human agency, otherwise it is questionable whether we are dealing with really human beings.

Philosophers are not renowned for their sensitivity to those contingencies of place that geographers take for granted. However, Peter Jones (1994: 167) recognizes that 'the distribution of resources across the world is entirely fortuitous and that it is morally unacceptable that people's lot in life should be determined by this accidental feature'. Barry (1989: 239) postulates Crusoe and Friday on two different islands, working equally hard and skilfully but with differences in production due to one island being fertile and the other barren, asserting that 'if anything can be called morally arbitrary – not reflecting any credit or discredit on the people concerned – it is this difference in the bounty of nature'. Unequal access to facilities such as good schools (e.g. Barry, 1989: 220–1), and fiscal disparities between local governments (e.g. Le Grand, 1991: 108, 128), are also part of the undeserved inheritance. Richard Miller explains that no one earns the right to be born to a family living in a spacious house in a New York suburb rather than on a straw mat in a Calcutta slum, yet this entails 'enormous differences in life prospects, given the same innate capacities and the same willingness to try' (Miller, 1992: 298).

The chance of birth in a particular place on the earth's highly uneven resource base carries no greater moral credit than being born to a rich or poor family, male or female, black or white. And such initial advantage as arises from the place of good fortune is readily transferred to future generations, similarly devoid of moral justification. In so far as right of access to unevenly distributed resources is constrained by the boundaries of nation-states, as accidents of history, then this source of inequality might also be considered morally irrelevant (Jones, 1994: 160). Indeed, the restrictive citizenship of Western liberal democracies has been described by Joseph Carens (1987: 252) as equivalent to inherited feudal privilege, and similarly hard to justify. He elaborates as follows:

> Citizenship in the modern world is a lot like feudal status in the medieval world. It is assigned at birth; for the most part it is not subject to change by the individual's will and efforts; and it has a major impact upon that person's life chances. To be born a citizen of an affluent country like Canada is like being born into the nobility (even though many belong to the lesser nobility). To be born a citizen of a poor country like Bangladesh is (for most) like being born into the peasantry in the Middle Ages. In this context, limiting entry to countries like Canada is a way of protecting a birthright privilege. (Carens, 1992: 26)

This is echoed by Veit Bader (1995: 214), who points out that citizenship laws combine criteria of birth or descent and territory: 'Morally no more defensible than all the other, like kinship, sex, age, region, residence, language, habits, culture, lifestyles, gender, religion, nationhood, social class, membership in churches, parties, and so on.'

Despite the rhetoric of globalization, most people still live in relatively closed worlds, 'trapped by the lottery of their birth' in their nation-state as a 'community of fate' (Hirst and Thompson, 1995). In Miller's apt phrasing, 'the charge of injustice points not to mere unchosen or undeserved disadvantages, but to imposed ones' (1992: 299). All this raises the question of whether the claim of sovereignty is anything more than a form of special pleading (Beitz, 1991: 246), designed to defend the privileged welfare of some territorially defined group. Philipe Van Parijs has given a Marxist twist to the interpretation of citizenship privilege, via the concept of exploitation:

> Like feudal exploitation, citizen exploitation pulls the distribution of income away from what it would be under pure market conditions, where only productive assets (wealth and skills) elicit differential rewards. The Marxist ethical imperative [to eliminate exploitation] requires that this form of exploitation too should be abolished. This generates at least a prima facie presumption in favour of anything that erodes the differential advantages attached to citizenship. (Van Parijs, 1992: 158)

Restrictions on border crossing are part of the structure responsible for the perpetuation of uneven development.

The conclusion is that inequalities in life chances among nation-states, as among individuals, cannot be defended morally, at least to their present extent. This leads to the proposition that the pursuit of greater (or complete) equality is a moral imperative. Whether the ultimate objective should be equality of opportunities, primary goods, capabilities or welfare outcomes, one way of achieving this might be to redistribute crucial resources. This is difficult enough in the usual philosopher's conception of individuals in a space-less world, and impossible if the resources in question are fixed attributes of the natural environment or built infrastructure. And, while financial capital is almost perfectly mobile in this age of electronic transactions, unencumbered international transfers in the form of aid from rich to poor countries have proved to be largely ineffectual in reducing inequalities among nation-states. The alternative is for people, and especially the poor, to move away from places of misfortune to those of greater advantage. Indeed, those liberals who extol the virtue of individual responsibility might be expected to applaud willingness to move. Hence, the case for open borders, as a matter of justice, as opposed to the closure of 'collective welfare chauvinism' (Bader, 1995: 215).

So, liberal theory 'has difficulty justifying boundary-keeping projects that disadvantage outsiders who want in, or insiders who want to get out' (Becker and Kymlicka, 1995: 466–7). The case is summarized as follows by Carens:

> Liberal egalitarianism entails a deep commitment to freedom of movement both as an important liberty in itself and a prerequisite for other freedoms. Thus, the presumption is for free migration and anyone who would defend restrictions faces

a heavy burden of proof. Nevertheless, restrictions may sometimes be justified because they will promote liberty and equality in the long run or because they are necessary to preserve a distinctive culture or way of life. (Carens, 1992: 25)

This formulation recognizes the intrinsic as well as the instrumental value of free movement, as something important enough to be argued as a fundamental human right, and also as a possible means to the ends of individual well-being and social justice. Carens's first proviso opens the way for strategic modifications of an unrestricted right of people to move and live where they chose, in the interests of the core values of liberal egalitarianism. The other proviso echoes Nett's earlier recognition that would-be migrants should be willing to obey local laws and respect local customs. It is on these considerations that the main practical and theoretical objections to open borders and free movement rest.

Libertarian and Communitarian Critiques of Liberal Egalitarianism

Rawls's liberal egalitarianism soon attracted criticism from the libertarian entitlement theory of Robert Nozick (1974). He argued that persons have the moral right to use such natural endowments as intelligence and skill to their advantage, providing that this does no harm to others. Similarly, persons are entitled to hold and benefit from natural resources, provided that they acquired them justly by initial acquisition or by transfer in the form of a gift, inheritance or purchase. His criterion for the justice of initial acquisition was that no other persons are thereby made worse off (Nozick, 1974: 178), a modification of John Locke's proviso that an individual is entitled to appropriate natural resources providing that there is as much and as good left in common for others.

Leaving aside the difficulty of demonstrating that no one is worse off as a consequence of particular private appropriations, and of tracing acquisition back historically, Nozick's coupling of natural endowments and acquired holdings raises the question of whether they may be different in some sense relevant to the place of good fortune. Jeffrey Reiman (1990: 173–5) suggests that the ownership of the external world can deprive others of resources whereas ownership of one's body cannot, though this still leaves scope for using personal attributes to advantage. A more pertinent difference is that people cannot change their entire body, but may be able to change their place on the earth's surface. However, Onora O'Neill (1991: 290) notes that the libertarian devotion to freedom does not extend to dismantling immigration laws: 'their stress on property rights entails an attrition of public space that eats into the freedom of movement and rights of abode of the unpropertied'. Libertarianism thus tends to support the proposition that groups of people are entitled to monopolize the resources of the territory which they occupy, as encouraged by the modern concepts of national citizenship and sovereignty: 'The nation provides its members with an inalienable collective property: the land in which they have the right to live their lives' (Poole, 1991: 96).

Michael Sandel (1982: 101) has questioned this proposition, arguing that 'for the community as a whole to deserve the natural assets in its province and the benefits that flow from them, it is necessary to assume that society has some pre-institutional status that individuals lack, for only in this way could the community be said to possess its assets in the strong, constitutive sense of possession necessary to a desert base'. And without this, to rephrase Sandel (1982: 92–3) in the individual context, no group anywhere deserves anything. A community, or nation, might claim an exclusive right to land in which they have mixed their labour, and even their blood, and to the advantages to be potentially derived. A similar argument might be applied to the infrastructure and social services, built up for the welfare of future generations as well as present people. But this still leaves unanswered the possible injustice of initial acquisition, and the moral arbitrariness of the good fortune of these people inheriting favourable conditions for sustaining a good life.

It is from the communitarianism espoused by Sandel and others that a more potent critique of the right to free movement has emerged. Communitarians recognize and value the existence of mutually interdependent groups of people, with common social practices and shared understandings that make up a distinctive culture or way of life. Their notion of the person situated or embedded in society, and defined at least in part by relationships with others, contrasts with the individual self-determination lauded by liberals. Communitarianism is therefore suspicious of the very idea of individual rights, such as to free movement. Indeed, leaving a community would be viewed as a rare rejection of its values. If rights beyond life and liberty are conceded, they are relative rather than universal. To Michael Walzer (1983: xv), 'these do not follow from our common humanity; they follow from shared conceptions of social goods; they are local and particular in character'.

Walzer's communitarianism values community not only in itself but also as a means towards an egalitarian form of justice. As he has famously asserted, the primary good that we distribute is membership in some human community: 'it is only as members somewhere that men and women can hope to share in all the other social goods – security, wealth, honor, office, and power – that communal life makes possible' (1983: 63). It is to protect the distinctive cultures of such communities, as a stable feature of human life, that open borders and free entry are opposed. 'If this distinctiveness is a value [...] then closure must be permitted somewhere' (1983: 39). And this closure may be made at the level of the sovereign state to protect the communities within. As summarized by Bader (1995: 213), Walzer's argument is that closure is necessary and legitimate to defend the shared meanings, values and ways of life of specific (ethnic, cultural, religious, linguistic, historical) political communities or states; for the production and development of collective political identity and attachment; and for the development of socially or culturally embedded, rich personalities.

As Bader points out, Walzer asserts the superimposition of ethnic, cultural and national identities over citizenship (see Walzer, 1995). The main point, in the context of borders and movement, is that Walzer defends a specific conception of the nation-state as a political community, which appears to be conflated with some aspects of an idealized local community. Critics of communitarianism point out that local cultures can incorporate inequalities based on status hierarchy and patriarchy, as well as being brutally repressive of difference and dissent. These are features from which some members might wish to escape, which Walzer permits, and which might be challenged and changed for the good by persons of other persuasion moving in. As at the local level, so at the scale of the nation-state, the value of preserving a distinctive way of life depends on what that way is. Chauvinistic cultural, ethnic, national or religious identities do not necessarily form the basis for a morally commendable state, as is illustrated by Nazi Germany or South Africa under apartheid.

It should be noted in passing that there are arguments for favouring co-nationals without a commitment to the 'associativism' central to communitarianism. These include not only the value of patriotism as a human virtue, debatable though this is, but also the importance of making formal rights of national citizenship effective, the relational nature of inequality which encourages equalization within nation-states, and the social nature of consumption which involves relative evaluations with close others such as neighbours (see Wellman, 2000; Coons, 2001). However, these points, along with opposing calls for spatially extensive beneficence (Corbridge, 1992; Smith, 1998), are more relevant to the re-distribution of resources than to population movement.

Whatever their limitations, the critiques of open borders and a right to free population movement require some response from liberal egalitarianism. The most obvious is that inward movement can rightly be restricted if there is a threat to the prevailing liberal order, as suggested by Carens (1992: 25). This might come in the form of an invading army or the infiltration of terrorists, both properly repulsed. The potential subversive who holds contrary values is a more difficult case for those dedicated to individual freedom, raising the tricky question of the place (if any) of illiberal groups in an otherwise liberal state committed to minority rights and multiculturalism (Smith, 2000b: 125–35). One may look no further than contemporary Britain to see that immigration of different others can elicit illiberal public and political reactions in a society which likes to consider itself liberal.

Liberal values and institutions may be threatened by some of the outcomes noted by Bader (1995: 215) and others in prudential arguments against unrestrained immigration. These include: the indigenous fear of being 'swamped' by outsiders; threats to public order; competition for jobs and rising unemployment; pressure on social security systems and public services; popular backlash towards newcomers, rising xenophobia and racism; and immigrants retreating into their own geographical and cultural enclaves. At the extreme, large-scale immigration could make most if not all the people concerned worse off, materially as well as with respect to loss of liberties, providing a utilitarian rationale for some degree of border closure.

Reviewing recent discussions in moral theory on free movement, Bader (1995: 213) favours a policy of 'fairly open borders'. Although he does not elaborate the point explicitly, there is a neat double entendre to his use of the word 'fairly'. One sense of the word is moderately or tolerably (as opposed to completely), implying borders open to a degree that reflects the kinds of practical consideration outlined above. The second sense suggests that the criteria adopted should be applied equitably. Equity requires that the extent of movement permitted should balance the welfare of both populations: immigrants/ emigrants and those who stay put. Equity also requires that differential treatment of individuals and groups is morally defensible. This raises some interesting issues, such as the morality of immigration policies favouring people with particular expertise or lots of money, rather than the desperately poor, at the expense of countries experiencing 'brain drain' and the like. Thus, Britain is actively recruiting social workers from South Africa, teachers from Eastern Europe, and doctors and nurses from wherever it can find them. This may be very much in the material interests of a society disinclined to raise public expenditure to train and pay enough service workers of its own, but at the morally dubious cost of exacerbating international inequalities. It may also be questioned whether the customary prioritization of 'refugees' is necessarily equitable. The United Nations definition of a refugee refers to a well-founded fear of persecution on grounds of race, religion and political opinion, yet it is far from self-evident that fear of dire poverty and starvation is a morally inferior motivation for migration.

Added to differential opening is the issue of directional opening. As was suggested in the introduction, there has been a shift of emphasis in recent years from the right to leave a country to the right of entry. The Universal Declaration of Human Rights treated the two quite differently (Barry and Goodin, 1992: 13). Article 13 gives everyone the right to freedom of movement and of residence within the borders of each state, and to leave any country, including their own, and to return to their own country. But with respect to entry, states retain the discretion to admit or refuse aliens, subject only to the right of individuals to seek and enjoy in other countries asylum from persecution. Dummett reacts as follows:

> Logically, it is an absurdity to assert a right of emigration without a complementary right of immigration unless there exist in fact (as in the mid-nineteenth century) a number of states which permit free entry. At present, no such state exists, and the right of emigration is not, and cannot be in these circumstances, a general human right exercisable in practice. (Dummett, 1992: 173)

There is an asymmetry of power between the individual and the state, with respect to both exit and entry, not dissimilar to that which makes welfare rights more difficult to achieve than liberty rights. The holder of a right to exit or enter (or to receive welfare) is the individual, but the reciprocal obligation is vested not in other individuals but in the state. Ultimately, even the most constraining state can grant exit in its own perceived interests, as in the case of some political dissidents in Eastern Europe under communism and in South Africa under apartheid. But the use of restrictions on exit as a means of social control tends to make these rare exceptions. The larger exit flows, actual or potential, are likely to be seeking escape from poverty or social insecurity rather than from political oppression. Expecting another state to recognize what may effectively be a welfare right for persons from beyond its borders brings us back to the more general moral issue of the spatial extent of beneficence, or responsibility to distant others in need.

So liberal egalitarianism is obliged to concede something of its commitment in principle to open borders and free population movement. The actual occupation of a particular territory by particular groups of people generates a sense of right to preferential benefit from its resources, which goes deeper than libertarian entitlement theory. Some restriction on entry follows. And even partial closure reinforces internal group identity and affiliation with 'our land', the defence of which becomes a supreme virtue. The modern, sovereign nation-state, for all its imperfections, has become the most reliable guarantor of personal security and welfare, however much this may vary among states. As Walzer (1995: 247) asserts, in response to Bader's criticism of his dedication to the nation-state as political community, 'It is not because of some historical misunderstanding that Jews, Armenians, Palestinians, Kurds, Estonians, and Tibetans lay claim and even fight for sovereign statehood', and that, if successful, they follow earlier precedents in the purpose of fostering and reproducing its cultural life. He recognizes that we live in an unsettled and mobile society yet claims that we are still 'creatures of community' with some common values, and with communal feelings and beliefs more stable than we once thought they were (Walzer, 1990: 11–18).

However, a nation-state of one's own is not the only answer to oppression. For example, if Palestinians had equal rights to Jews in a truly liberal (and multicultural) state of Israel, their national territorial aspirations might carry less weight. And to argue that the cultural integrity of local, far less national, entities must always trump the rights of outsiders to enhanced security and welfare goes too far. Subject to some prudential constraints on volume of immigration, to protect aspects of national life that are valuable as well as distinctive, international population movements on a larger scale than presently permitted seem warranted on grounds of both liberty and justice. Bader's notion of 'fairly open borders', with its element of ambiguity and indeterminacy, captures the sense of direction, if not the complete consensus, of contemporary moral argument.

Conclusion

We are all descendants of immigrants. [...] The fact that we are born with legs and intelligence opens to us ever new spatial and intellectual horizons. [...] The human ability to migrate has been one of our greatest assets of survival, allowing us to free ourselves of geographic constraints, from bondage to the earth. Bosnians, East European refugees in German hostels, Chinese, Haitian, and Vietnamese refugees, like our ancestors and ourselves, whoever we may be, are searching for their home. (Tucker, 1994: 186)

[R]adical international redistribution is the first and most important option in the struggle against structural poverty and inequality. To the degree that this policy does not succeed, one of the most important causes of forced migration cannot be removed and the legitimacy of all possible other normative arguments in favor of closed borders will be severely undermined as a consequence. [...] Closure under conditions of 'rough' equality differs radically from closure under conditions of systematic exploitation, oppression, discrimination and exclusion. (Bader, 1995: 216–19)

One of the revolutions of modernity was potentially to release people from the confines of local resources, opportunities and moral conventions. While the pre-modern era saw some population migrations of spectacular distance and scale, it was the European making of America and colonization of other sub-continents that so greatly influenced the political shape of the contemporary world. Now the freedom that generations of nineteenth-century Europeans took for granted has to be argued in the language of individual human rights, in the face of the exclusionary power of sovereign states. Paradoxically, some question the continuing salience of the very concept of rights (e.g. Low and Gleeson, 1999), in general and as applied to population movement. Chris Brown, writing from a Marxian perspective, expresses these reservations:

To think of the ethics of migration in the context of the individual it is necessary to think of the latter as an independent maker of choices about his or her future; it can then be asked whether the individual has the right to make certain choices, what sort of networks of obligations and duties surround the making of such choices, and so on. [...] It makes little sense to ask whether states have the right to restrict entry or prevent departures, to 'poach' skilled workers and profession-als from elsewhere, or to stop similar people from leaving. Such matters are in the last resort a function of a capitalist economy and to ask whether states have a right to behave in this way is like asking whether capitalists have a right to make a profit. (Brown, 1992: 134)

The assertion of rights has nevertheless been a powerful if uneven force for human betterment in recent history. And the continuing inclination to specify rights in legal form, as in post-apartheid South Africa, has widened opportuni-ties to complete the conversion from moral rights to *de jure* rights, and to rights actually implemented and experienced. But this tends to be within the context of existing nation-states, with universal rights as elusive as ever. If it does nothing else, the language of rights enables people to express aspirations, arising from unsatisfactory background conditions that may appear unyielding to challenge and change.

The current significance of debates about a right to free population movement reflects above all the extent of spatial disparities in life chances, internationally and within nation-states. Writers from a liberal position (like Carens and Bader) share with Marxists the stress on international redistribution, even if they differ on the extent of institutional reform (or revolution) required to achieve it. The first priority is to radically change the background conditions that impel people to leave their home and homeland, and perhaps even their family, to risk death on some rickety boat or barbed-wire fence for the sake of a better life. For them, such concepts as multi-layered citizenship in newly emerging 'spaces of democracy' mean little without basic need satisfaction: some assurance of physical survival – preferably in a situation guaranteeing elementary civil rights such as freedom from fear of arbitrary arrest, torture and extra-judicial execution.

Politicians promote a New World Order, driven by the advance of electronic communications and commerce. But their rhetoric promises little to the majority of inhabitants of the lands of origin of the major potential migration streams, from which freedom of movement itself will continue to be unequally distributed. International markets for persons with certain talents are trumpeted, while fortress-like defences are erected against others. Some predictions of winners and losers in the coming world order have been made by the foundation head of the European Bank for Reconstruction and Development, Jacques Attali (1991: 5–6), who envisages a future of 'rich nomads' and 'poor nomads'. The rich, as the consumer citizens of the world's privileged regions, will be able to participate in the liberal market culture of political and economic choice, roaming the planet in search of the information, sensations and goods that only they can afford, while yearning for lost human fellowship, home and community. They will be confronted by the poor, roving masses of boat people on a planetary scale, seeking escape from the destitute periphery where most of the world's population will continue to live. Their encounters will be a far cry from the rich cultural diversity of the porous places sometimes celebrated in contemporary geography.

However, the reality is likely to be a more spatially fixed polarization, with increasing extremes of rich and poor juxtaposed at different geographical scales. That their interface will be policed more assiduously than ever, from penetration by asylum seekers, economic migrants and potential terrorists, is the direction of recent experience. No matter how persuasive philosophical argument might be, anything approaching free population movement still appears to be a right we are not ready for. And as long as this is so, defending the boundaries of democratic participation and other associated privileges will continue to challenge the ethics as well as the politics of liberalism.

Note

Parts of this chapter draw on earlier publications, the research for which was supported by a Leverhulme Fellowship. Thanks to Moritz Lennert for some helpful suggestions.

References

Attali, J. (1991) *Millennium: Winners and Losers in the Coming World Order*. New York: Times Books.

Bader, V. (1995) 'Radical democracy, community, and justice: or, what is wrong with communitarianism?', *Political Theory*, 23, 211–46.

Baker, J. (1987) *Arguing for Equality*. London: Verso.

Barry, B. (1989) *Theories of Justice (A Treatise on Social Justice, Volume 1)*. London: Harvester Wheatsheaf.

Barry, B. and Goodin, R. E. (eds) (1992) *Free Movement: Ethical Issues in the Transnational Migration of People and of Money*. Hemel Hempstead: Harvester Wheatsheaf.

Becker, L. and Kymlicka, W. (1995) 'Introduction: symposium on citizenship, democracy and education', *Ethics*, 105, 465–7.

Beitz, C. R. (1991) 'Sovereignty and morality in international affairs', in D. Held (ed.), *Political Theory Today*. Cambridge: Polity Press, pp. 236–54.

Benton, C. (1995) *A Different World: Emigré Architects in Britain 1928–1958*. London: RIBA Heinz Gallery.

Brown, C. (1992) 'Marxism and the transnational migration of people: ethical issues', in B. Barry and R. E. Goodin (eds), *Free Movement: Ethical Issues in the Transnational Migration of People and of Money*. Hemel Hempstead: Harvester Wheatsheaf, pp. 127–54.

Carens, J. H. (1987) 'Aliens and citizens: the case for open borders', *The Review of Politics*, 49, 251–73.

Carens, J. H. (1992) 'Migration and morality: a liberal egalitarian perspective', in B. Barry and R. E. Goodin (eds), *Free Movement: Ethical Issues in the Transnational Migration of People and of Money*. Hemel Hempstead: Harvester Wheatsheaf, pp. 25–47.

Coons, C. (2001) 'Wellman's "reductive" justification for redistributive policies that favor compatriots', *Ethics*, 111, 782–8.

Corbridge, S. (1993) 'Marxisms, modernities and moralities: development praxis and the claims of distant strangers', *Environment and Planning D: Society and Space*, 11, 449–72.

Dummett, A. (1992) 'Natural law, solidarity and international justice', in B. Barry and R. E. Goodin (eds), *Free Movement: Ethical Issues in the Transnational Migration of People and of Money*. Hemel Hempstead: Harvester Wheatsheaf, pp. 169–80.

Hirst, P. and Thompson, G. (1995) 'Globalization and the future of the nation state', *Economy and Society*, 24, 408–42.

Jones, P. (1994) *Rights*. London: Macmillan.

Le Grand, J. (1991) *Ethics and Choice: An Essay in Economics and Applied Philosophy*. London: Harper Collins.

Low, N. and Gleeson, B. (1999) 'Geography, justice and the limits of rights', in J. D. Proctor and D. M. Smith (eds), *Geography and Ethics: Journeys in a Moral Terrain*. London: Routledge, pp. 30–43.

Miller, R. W. (1992) *Moral Differences: Truth, Justice and Conscience in a World of Conflict*. Princeton, NJ: Princeton University Press.

Nett, R. (1971) 'The civil right we are not ready for: the right of free movement of people on the face of the earth', *Ethics*, 81, 212–27.

Nozick, R. (1974) *Anarchy, State, and Utopia*. New York: Basic Books.

O'Neill, O. (1991) 'Transnational justice', in D. Held (ed.), *Political Theory Today*. Cambridge: Polity Press, pp. 277–304.

Poole, R. (1991) *Morality and Modernity*. London: Routledge.

Proctor, J. D. and Smith, D. M. (eds) (1999) *Geography and Ethics: Journeys in a Moral Terrain*. London: Routledge.

Rawls, J. (1971) *A Theory of Justice*. Cambridge, MA: Harvard University Press.

Reiman, J. (1990) *Justice and Modern Moral Philosophy*. New Haven, CT: Yale University Press.

Sandel, M. (1982) *Liberalism and the Limits of Justice*. Cambridge: Cambridge University Press.

Shapiro, M. J. (1994) 'Moral geographies and the ethics of post-sovereignty', *Public Culture*, 6, 479–502.

Smith, D. M. (1998) 'How far should we care? On the spatial scope of beneficence', *Progress in Human Geography*, 22, 15–38.

Smith, D. M. (2000a) 'Moral progress in human geography: transcending the place of good fortune', *Progress in Human Geography*, 24, 1–18.

Smith, D. M. (2000b) *Moral Geographies: Ethics in a World of Difference*. Edinburgh: Edinburgh University Press.

Tucker, A. (1994) 'In search of home', *Journal of Applied Philosophy*, 11, 181–7.

Van Parijs, P. (1992) 'Commentary: citizenship exploitation, unequal exchange and the breakdown of popular sovereignty', in B. Barry and R. E. Goodin (eds), *Free Movement: Ethical Issues in the Transnational Migration of People and of Money*. Hemel Hempstead: Harvester Wheatsheaf, pp. 155–65.

Walzer, M. (1983) *Spheres of Justice: A Defence of Pluralism and Equality*. Oxford: Basil Blackwell.

Walzer, M. (1990) 'The communitarian critique of liberalism', *Political Theory*, 18, 6–23.

Walzer, M. (1995) 'Response to Veit Bader', *Political Theory*, 23, 247–9.

Wellman, C. H. (2000) 'Relational facts in liberal political theory: is there magic in the pronoun "my"?', *Ethics*, 110, 537–62.

Woodward, J. (1992) 'Commentary: liberalism and migration', in B. Barry and R. E. Goodin (eds), *Free Movement: Ethical Issues in the Transnational Migration of People and of Money*. Hemel Hempstead: Harvester Wheatsheaf, pp. 59–84.

7 Cities as Spaces of Democracy: Complexity, Scale and Governance

Murray Low

It is easy to view cities as actual or potential spaces of democracy. Cities are central to the everyday experiences of much of the world's population, and are also favoured objects of study that connect geography with other social science disciplines. Moreover, the city has a particular place in Western thought about democracy. This is because of the iconic status of classical Greek, medieval Italian and other city-states in debates about the origins and destinations of democratic practices. This association between cities and democracy has been reinforced by affinities between the city and the idea of citizenship; by ongoing traditions emphasizing the need for more participatory, local forms of democratic governance; and by recent concerns with social capital as a means of 'making democracy work' (Putnam, 1993). This chapter will argue that, in fact, cities are not good models for democracy in general, and that it is hazardous to view them as uniquely important sites for deepening democratic governance.

There are two important senses in which the spatiality of the city is thought to be important to debates about democracy. First, cities are favoured spaces for thinking about democracy because of the importance attributed to the types of democratic practice that they have made or can make possible. In particular, the interpersonal proximity and density of contact facilitated by cities have always held out the hope of cities as places where better, more participatory, or at least more involving democratic practices might thrive. This hope is connected to interpretations of democracy's past in which city-states are figured as privileged sites for the exploration of forms of democracy seen as more theoretically defensible or practically engaging than modern representative institutions operating in cities and at broader scales.

Secondly, cities can be seen as crucial spaces for democracy because of the scope and roles of their governments in relation to particular geo-historical transformations. Key transformations in political institutions have often involved changes in the scale organization of politics. In narratives about modernity, the construction of 'national' or territorial states occurred by reorganizing a prior political landscape defined by the often mosaic-like territorial structures of landownership and networks of cities. In many accounts (e.g. Dahl, 1989; Held,

1995; Manin, 1997), the transition from democratic city-states to nation-states made new forms of democratic practice necessary. In these stories, transformations in the scale of political institutions led to altered views about the definition of democracy, about what sorts of institution embody democracy, and about what we can expect of democratic rule. As noted in the introduction, this spatial story is not a novelty: it was already characteristic of reflections on democracy and modernity around the turn of the nineteenth century, in the writings, for example, of James Madison, Benjamin Constant and Alexis de Tocqueville. The current relevance of the narrative concerns its continuation in a post- or less-national world (Taylor, 1995). In particular, recent discussions of globalization have placed a considerable theoretical and practical-political emphasis on cities and city-regions as key protagonists in a post-national global order. A stronger focus on the possibilities presented by cities for the deepening of democracy has ensued (Borja and Castells, 1997; Isin, 2000).

Arguments about the potential political role of cities in an emergent post-national order often draw on both of these lines of thinking. Normative claims about the importance of proximity in constructing more democratic relations between citizens, and between citizens and government, can be used to support arguments about global transformations leading to inevitable changes in the scale of politics and democracy. As a result, cities' increasing responsibilities as spaces of democracy can be presented as both necessary and desirable (Borja and Castells, 1997). In this chapter, I explore three issues concerning cities as political spaces that connect with both of the senses noted above in which cities are ascribed a special status when it comes to thinking about democratic practices. First, I consider the issue of the complexity of urban political space, to put in perspective claims about the privileged interaction-contexts cities may provide for democracy to flourish. Secondly, the connections between cities and democracy raise questions about the ordering of relations between governing organizations at different spatial scales. These relations are often viewed as external to the scope of democracy as we understand it. I suggest that a constructive sense of democracy's potential should, in some way, internalize these relations to democracy. This has implications for how we think about what democracy involves and also, by relativizing cities within broader geopolitical networks, makes them seem less plausible as privileged sites of democracy in themselves. Thirdly, most current discussions of urban politics emphasize the ongoing pluralization of the political agents involved in running cities. This is usually discussed under the heading of new governance arrangements. Some of these agents (e.g. communities, unions, neighbourhood movements, voluntary sector organizations) can fit into some existing theories of democracy. The involvement of other actors, of businesses in particular, seems harder to square with our inherited sense of democracy as the rule of the people. Nonetheless, I will argue that, in spite of the complications involved in relating democracy and governance, this theme does have the merit of forcing us to view the particular practices actually or potentially comprising democracy in cities in more open and productive ways.

Each of these three issues presents us with many difficulties, and it is not possible to resolve them here. They complicate many of the more straightforward connections made between urban spaces and democratic spaces. But in doing so they open up space for imagining new ways of reconciling the heterogeneity and complexity of cities with the need for accountability. Democracy, like cities, has to be thought of as complicated and capable of assuming many forms in different contexts. Simple ideas about social interaction and proximity, local autonomy and government as opposed to governance are impediments to this task.

Cities, Complexity and Scale

The privilege accorded cities as sites for democracy often rests on a set of microsociological foundations. This involves the simplification of the city into a political space whose occupants can face each other, encounter each other on the street, gather together, use their senses to negotiate their relationships with one another, struggle with one another, and so on (see Low, 1997). It can also involve the figuring of the city as a space where anonymity, non-encounters and isolation are characteristic general conditions to which building institutions facilitating denser interaction are an appropriate response. Both of these lines of thought depend on the idea that cities do, or should, produce new kinds of social relationship through the heightened interpersonal proximity they make possible. There have been many critical discussions of such urban imaginings (e.g. Castells, 1977; Sennett, 1977; Boden and Molotch, 1994; Amin and Thrift, 2002). Yet, when not themselves subjects of reflection, generalizations about urban interaction or non-interaction underpin a remarkable amount of work on the city. They have played a central role in thinking about the importance of cities as sites of cultural or industrial activity (e.g. Scott, 1988; Zukin, 1995), as well as in normative accounts of democracy and social justice drawing on models of urban life (Young, 1990).

Why does this matter for thinking about cities as democratic spaces? The characterization of urban life in terms of the reality or possibility of dense, proximate, interpersonal interactions – both between citizens and between citizens and agents of government – is an important element in much thinking about democracy. The models of democracy associated with the city-states of the past, or with the Paris Commune, or with contemporary Barcelona or Porto Allegre, are all informed by the idea that proximity between urban citizens, and between them and their governments, makes possible more direct, knowledgeable, interactive forms of democracy than those associated with polities organized at broader scales. Moreover, claims concerning the benefits of democracy at the urban scale, in terms of the proximity of government to the people and of the direct involvement of citizens, are common in everyday city politics around the world.

The case of London is quite instructive. Like other recent innovations in local government in the UK, the introduction of a successor to the Greater London Council, which was abolished in the 1980s, was justified in terms of bringing government closer to residents, a justification often expressed in terms of bodies and interaction (Labour Party (UK) 1996; UK Government, 2001). As is common in contemporary discussions of localized democracy, the metaphor of democracy as a form of 'voice' – emphasizing the directness of the flow of communication from citizens through the new institutions – was frequently mobilized to legitimate the new Authority. Crucially, the new arrangements were seen as a way of bypassing institutions that interpose themselves between the London electorate and those who govern them. The elected assembly element of the UK model of local councils was downgraded, and a directly elected executive Mayor was introduced. The Mayor would be able to personify and 'speak up' for the city, giving it 'a voice'. There was much commentary about the way in which this arrangement would encourage a post-party politics in which one highly visible leader would improve upon the more anonymous workings of normal politics by putting Londoners 'in touch' with their government (e.g. Phillips, 1997; Jenkins, 1998a). In the event, this perception was reinforced by the election of an independent as the first Mayor in 2001. A range of interactive opportunities were on the agenda, including town meetings, various types of forum related to specific areas of policy, and Mayor's question times. Each policy-making mechanism specified in the Act that established the Greater London Authority was accompanied by a host of specifications about the involvement of communities, businesses, and other specified social interests. The new headquarters of the Authority are a glass lens-shaped building meant to symbolize the abolition of distance between Londoners and power, and a new transparency of democratic political life.

Now, there is nothing intrinsically wrong with or un-democratic about having an elected Authority at the scale of Greater London. Arguments for the Authority in terms of both community identification with London and better overall strategic management of the city have much plausibility. Yet the legitimating rhetoric surrounding the Authority was at best misleading, and at worst set up impossible expectations as to what it could achieve in democratic terms. London is obviously a highly complex urban centre, whose inhabitants cannot speak with one voice though one leader without a great deal of selection and aggregation going on. Nor can Londoners be routinely involved *en masse* in policy-making and town meetings. The Greater London Authority is not necessarily more 'transparent' than governments at broader scales, and no 'closer' to London's electorates than the national government in Westminster and Whitehall. London's residents (or at least those of them who had political citizenship rights) were not in any straightforward sense 'voiceless' or unrepresented before its advent. They did and still do elect representatives to the national parliament and to borough councils (which, as opponents of the creation of the GLA were quick to point, were certainly 'closer' to the people than the new institutions would be).

The next section will consider the important issue of the relationship between urban autonomy and democracy. Here it is enough to note the use of ideas about proximity and directness in the London case. This example reminds us that there is a strong ideological dimension to microsociologically grounded understandings of why cities matter in democratic terms. Boden and Molotch (1994) present a careful and nuanced set of arguments defending the importance of face-to-face interaction as a form of interactive communication, but these imply that there is a low spatial threshold where the benefits of proximity lose much of their purchase. The idealized microsociologies of democracy typically depend on simplified imaginary geographies that emphasize generalizations about encounters and everyday life in the city. Ideologically compelling as these can be, as overall characterizations they are not very helpful. Cities are far more complex spaces than such microsociological accounts suggest. They are characterized by various different forms of socially structured interaction and non-interaction; by social relationships embodying different degrees of knowledge and ignorance; and, crucially, by contacts and communications mediated by an array of technologies, institutions and organizations.

One way out of these difficulties is to say that cities are characterized by social complexity, and to try to evaluate their potential for democracy on this basis. By itself, this is a risky strategy, and the difficulties are perhaps one reason why simple interactive stories about co-present urban residents and government retain so much force in theory and in the world. Stories about 'complex societies' have played a key role in arguments about the possible shape democracy can take in contemporary conditions. Robert Dahl's arguments about pluralist politics in New Haven, Connecticut, are embedded in a narrative about modernity and social differentiation where social differentiation enables but also constrains what we can expect from modern urban democracy (see Dahl, 1960; Judge, 1995). Social complexity can also provide fuel for arguments that democracy is a highly problematic form of rule under contemporary conditions. Hayek's (1960) arguments about the need to limit the scope of democracy in the face of the incalculably complex and uncontrollable market processes that characterize modern societies comes to mind (see also Zolo, 1990).

To argue that cities are complex or differentiated does not, however, get us very far in thinking through the possibilities for urban democracy. By itself, it is just as unhelpful as imagining cities as built up from microsociological relations founded on proximity. Cities are social spaces that are differentiated or complex in specific ways, depending on their social, cultural and economic histories, position in the urban hierarchy, and governmental practices. Noting that Mumbai, New York, Recife, Athens and Beijing are all complex is an important baseline for thinking about democracy in these contexts, and does act as a corrective to nostalgia about city-states or community. Moreover, it suggests that we should not view the politics of the city as necessarily more or less complex than that of entities defined at other scales. City politics is not, in a mediatized age, necessarily more visible or transparent than that of the nation-state or even of international organizations. Outside the focused conduct of highly localized political discussion and negotiation, its inhabitants are not necessarily any more able to participate directly in its politics than they are at other scales. But the heuristic usefulness of the idea of complexity does not provide much analytic purchase on democracy and democratization in different cities. For that, more focused analysis and evaluation of different institutions and procedures in terms of their communicative potential, mechanisms of accountability, and potentials for change in differently complex urban environments is necessary. Amin and Thrift (2002: 157) have emphasized that the city, far from being reducible to a sociology of interaction and proximity, is 'brimful of different kinds of political space'. I agree with them that cities are interesting and important 'settings for the *practice* of democracy' as opposed to 'formative of a particular *form* of democracy' (Amin and Thrift, 2002: 152) that would be identifiably urban and superior to other hypothetical forms at other scales. Following through on this understanding requires more precision about the qualities of political institutions that link up (or fail to link up) different kinds of space, and the manner in which these more or less effectively mediate the different complexities making up differently situated cities.

Democracy, Autonomy and City Limits

It is difficult to imagine an operative democracy where governments are not sufficiently autonomous of spatially external influences and pressures to respond unequivocally to 'the people'. In reality, however, for all governments professing democratic credentials, the degree to which their actions are determined endogenously by the preferences of their peoples is highly variable across policy areas. Moreover, because of globalization and similar processes, this problem for democracy could be said to be increasing in intensity. Human geographers and others have for some time been exploring the relationships between political institutions organized at different geographical scales. Topics of debate have included transnational processes; the geopolitics of the world system; emergent forms of potentially post-national politics such as the European Union or trans-border regions; and relationships between cities and regions, on the one hand, and the national and international political institutions in which they are embedded on the other.

We can speak here of 'democracy's edges' (Shapiro and Hacker-Cordon, 1999). More apposite in the urban case would be Peterson's (1981) metaphor of 'city limits', developed in the context of a critique of cherished ideas, sometimes reinforced by the case-by-case nature of much urban political research, about local or urban political autonomy. A wedding of ideas about local autonomy with ideas about urban democracy has been important in recent discussions of re-scaling political life, particularly in the case of those seeing globalization as an opportunity for shaping more positive forms of city-centred politics. In recent decades, decentralization has been seen as a political good by many on both the right and left of the political spectrum, although many of the technical arguments often made for it turn out to be weaker than might be supposed (Rodriguez-Pose and Gill, 2003). In addition, even proponents of a greater role for cities or city-regions are usually aware of the gap between urban/regional capacities and resources and the larger roles that the 'hollowing out' of the nation-state might make it possible or necessary for them to take on. The strongest grounds for greater urban autonomy are often normative ones, depending on the strong democratic resonances of the idea that autonomy implies self-determination. Normative arguments for local autonomy are connected to discourses about cities facilitating democracy because of their potentially interactive character. The key reason for re-centring politics on cities in a context of globalization often turn out to have to do with proximity relations (Borja and Castells, 1997).

Autonomy is usually said to be a key democratic value. Yet, some of the most authoritarian forms of political power over the last century were firmly grounded in the promotion of autonomy at various scales. Nationalist discourses in particular have invoked notions of autonomy to justify reactionary, organicist and non-democratic concepts of popular representation. The defence of local privilege, institutions, and practices from external interference characterized resistance to democratization and universal citizenship in many contexts in the late twentieth century, for example, in the cities and states of the American South. By itself, local autonomy is not a guarantee of satisfactory democratic decision-making procedures, though it has been associated with the development of viable democratic procedures and practices in cities around the world. There have been interesting attempts to define and categorize degrees and types of local autonomy on the basis of different political and legal systems (see e.g. Clark, 1985; Wolman, 1995), which suggest that local autonomy may be most productively thought about as a property whose meaning varies between different contexts. Yet, given the role of local autonomy in supporting arguments about the desirable roles of cities in a globalizing world, it is a little surprising that its specific political-legal implications and its relationship to democracy have not received more attention. This is all the more so, given the fact that the same processes of geopolitical reorganization which form the context for arguments in favour of devolution and the rebirth of city-states have provoked a flow of critical discussion about the concept of sovereignty, a closely related concept to autonomy.

The policy-making autonomy of cities, then, is not necessarily connected to their democratic credentials. It might even be argued that running cities democratically necessarily involves extensive interaction with and even management by democratic government organizations at other scales. Limitations and diminutions of urban political autonomy may therefore be part and parcel of democratization processes. Legalès (2002) develops a set of arguments directed at what he sees as mythical understandings of the political pasts and potential futures of European cities as autonomous actors. He suggests that we view cities as 'incomplete societies', able to some degree to project themselves as autonomous agents, but only partially, because of their internal divisions, conflicts, specific socio-economic profiles, and their openness to outside relationships, influences, resources and institutions. His arguments are suggestive in relation to dilemmas about cities, democracy and political scale. Citizens involved in city politics make up partial citizenries, simultaneously involved in broader constituencies with a wider geographical reach and interest (including the policy choices of other cities) as well as being involved with jurisdictions of more limited scope within the city itself.

They also have interests in the actions of national and regional governments, and when these interfere in or structure politics in urban centres it is far from clear that a basic democratic norm is being in some sense violated.

To some extent, the problem of defining different 'peoples' or constituencies that lies at the heart of the intuition that intergovernmental relations, in particular, are in some sense necessarily external to democracy is a false one (Hurley, 1999). Democracy does not involve processes of political communication between tightly confined political communities and their governments, a vision which at both local and national levels can give rise to inward-looking, static and traditionalist notions of who 'the people' involved in democracy are. In reality, it is a messy affair of negotiation between multiple and overlapping popular constituencies at different geopolitical scales, in which dynamic, contested and shifting definitions of popular interests, needs and identities are necessarily at stake. Thinking about democracy in urban contexts might therefore require thinking about the specifically democratic, as opposed to autonomy-limiting, character of the intergovernmental links, institutional connections, flows of information and influence and the mediation of competing interests at different scales that cities inevitably face. The interventions, for example, of the national Labour government to limit the tasks facing the Greater London Authority, particularly in the areas of transport policy and revenue, and in the politics of the first Mayoral election, only seem illegitimate, as opposed to unwise or contestable, in the context of an equation between urban democracy and urban autonomy propagated by the same government in setting up the Authority in the first place (Jenkins, 1998b). The merits of the government's actions certainly deserved all the scrutiny they received, but since London is a focus of concern and interest for citizens distributed well beyond the perimeter of the Authority's jurisdiction, it would be difficult to construe them as un-democratic were it not for the prevalence of this equation between democracy and autonomy.

These problems are not new ones, even if they are more sharply posed by recent phenomena such as globalization and the evolution of the European Union. Democratization has historically involved the formation of mediating institutions such as political parties, labour unions, social movements, and local or urban government associations, all of which link constituencies across differentiated jurisdictional spaces, including beyond the national scale. It is remarkable how little discussion these organizations receive in connection with urban democracy (as opposed to urban politics or governance), even though they are crucial to the insertion of cities across the world in the geographical complexities of contemporary democracy. For example, Borja and Castells (1997) discuss the importance of the formation of urban-based political parties as a step towards revitalizing urban citizenship. Yet the creativity of urban democracy is often rooted in the particular tensions channelled through cities' connections with constituencies at broader scales through party systems and intergovernmental links. The case of Barcelona comes to mind, where rival yet connected versions of nationalism have been invented and played out through the relationship of city to province. So too does that of New York City, with its complex history of political innovation rooted in tensions, mediated by complex national party organizations, between city, state and federal governments (Pecorella, 1987; Shefter, 1996).

When we consider cities in connection with intergovernmental relations we are in a zone where it can be hard for inherited ideas about democracy to make sense. Intergovernmental and more broadly interscalar relations have often been viewed as interstitial to workable ideas about democratic politics. Yet, consideration of unavoidable intergovernmental elements in the operations of democratic city politics forces us to abandon any simplistic picture of autonomous democratic cities where politics and policies somehow derive from the internal transmission of urban constituencies' wants, preferences or needs. Cities are incomplete societies with incomplete citizenries and cannot be viewed as democratic entities in and of themselves. The response cannot be, however, to resort to equally unhelpful ideas about the necessarily greater democratic legitimacy inhering in governmental organizations at larger scales, as though there were a simple positive relationship between constituency size and legitimacy. This would simply reproduce the same kind of picture at a different level and present an equally essentialist picture of the ongoing primacy of national democracy. The fact is that intergovernmental relations are systematically relativizing for all the different scales 'in' which democracy is taking place: none of these is best thought of as fundamental for how democracy works. Nonetheless, these connections between constituencies and jurisdications are also *constitutive of* democracy. It is perhaps natural in studying cities that researchers and theorists are predisposed to scrutinize the relationship between urban society and urban government. But where the relationship between models of democracy and the real-world geography of political institutions is at stake, rather than asserting the necessary priority of one space of democracy over others, a greater emphasis should probably be given to evaluating the means through which partial popular constituencies interact, negotiate and inter-organize to invent ways to make democracy work.

Governance and urban democracy

The third theme I want to consider is the de-centring of the contemporary institutions of urban governance. It is a common theme in urban studies that *government* – understood as a relationship between state and society – has been replaced by more complex *governance* arrangements. In the latter, different networks of public and private actors come together, or are brought together, to formulate and implement solutions to public problems. These actors generally include elements of central and local governments, businesses, as well as specialized public forums, neighbourhood organizations, voluntary sector organizations and other NGOs.

'Government' is at least associated with the possibility of democracy, in that a clear flow of influence from the people to something called government capable of implementing this will forms the basis of much democratic theory. 'Government' arrangements allow us to imagine the channelling of popular influence towards an agent of that influence that is clearly identifiable and hence potentially accountable to the public. 'Governance' scrambles these clear flows of influence and implementation and obscures loci of responsibility. Unsurprisingly, a key question in much of the literature on governance is the degree to which these sorts of specialized, multi-actor arrangements represent a shift away from democratic urban management (Burns, 2000). As the emergence of non- or less- state forms of governance proliferate (from subcontracting, private finance arrangements, through regeneration partnerships to area-based management schemes in which the state may cede a wide range of powers to private interests in large tracts of cities), conventional liberal forms of democracy, in which publicly accountable politicians are periodically elected and have responsibility for the management of cities, seem to have less and less purchase on what is really going on. What seem like accumulating experiments in the organizational form of public policy-making and implementation could clearly shade, at some point that is difficult to determine, into an urban democratic deficit of alarming proportions.

'New forms of governance' is a topic about which normative argument is quite difficult. As with globalization, or modernization, it is hard to argue with those who write about it as a clear, and irreversible, shift in the way the political world works. Critics (e.g. Jessop, 2001) argue that a transition to new mechanisms of governance is in some sense inevitable, and focus on working out its contradictions and dynamics as a form of rule. On the other hand, enthusiasts imply that there is something ideologically backward in expressing concern about these kinds of change (Goss, 2001). Nonetheless, 'move on' might well work as a slogan for both of these positions. Of course, blanket criticism on democratic or other grounds seems very difficult to sustain. 'Governance' in the urban context means a number of different things, and its promotion can be conducted within a range of different ideological frames that have quite different implications for understanding the ongoing role of the state in urban life (Legalès, 2002). The label covers so many policy-making arrangements that it is not clear how to begin except on a case-by-case basis. The real extent of governance–type innovations, and their variability and uneven development across national contexts, also creates difficulties in evaluating the degree to which a shift from government to governance is a generalized problem facing democratic theory (John, 2001).

New governance arrangements confront democratic theory with three general problems. The first problem concerns the direct involvement of business in making, implementing and managing urban policy. This is different from the problem raised by the perennial question in urban studies about the disproportionate power and influence of certain businesses over city policy-making. It is a problem raised by the formal incorporation of business into the running of cities. While in the first case, the activities and power of business in city politics are normatively suspect within the very terms used by urban government to legitimate itself, in the second case, of new urban governance, it is not. This transition has been mirrored to some degree in academic writing on city politics. Formerly, arguments that businesses had disproportionate power over urban political agendas and decisions were characteristically countered by suggestions that this could not be shown to be systematically the case. This, generally speaking, was an important dimension of the 'pluralist' analysis of city politics in the 1960s and 1970s (Dahl, 1960; Polsby, 1980; Judge, 1995). Now, it is more common to suggest that business power, exercised more or less informally through structures such as 'urban regimes' (Stone, 1993) or through the various forms of partnership organization characteristic of governance, is at least a necessary evil, and perhaps justifiable on grounds of efficiency, effectiveness and flexibility.

Over time, democracy has been understood through its differences from other forms of rule, not in terms of an essential, easily grasped meaning. In Western political thought, comparisons with monarchy and oligarchy, or more recently with dictatorship, have been important. In the urban context, the contrast between democracy and rule by elites has been equally central. Operationalizing the concept of elites in cities has not been straightforward, as different kinds and combinations of power-resource may underpin the ability of certain social actors to influence policy and political institutions more than citizens at large. But whatever the nuances, it is hard to avoid the role that business power has played in helping to define (negatively in most accounts) what democratic cities might be like. In seeming to ratify a key role for businesses in running cities, new governance arrangements do therefore seem to move across a crucial boundary that defines urban democracy. There are different visions of democracy, and some of the new governance arrangements are compatible with those that endorse a shift to processes of negotiation and consensus-building among groups or associations, including businesses (Amin, 1997). Here, it is plausible that some of the structural leverage of powerful economic actors is tamed through dialogic processes, counterbalancing their presence with that of other interests and citizen representation. On paper, at any rate, governance arrangements often look like structured exercises in networked pluralism, rather than elitism, in so far as in the former case the participation of, and negotiation between, different interest groups or stakeholders are designed into urban management. It might even be argued that, despite the risk of capture of these arrangements by business, the outlook of businesses in forums, partnerships and other instruments of governance, has transformative potential because it engages them in structured relationships with government and other city groups. However, despite the fact that democracy in capitalist societies has always involved a degree of compromise with business interests, unlike the case of intergovernmental relations, this structuring limitation is not internal to democracy. The actual or potential use of power by business (based in financial, cultural, knowledge and social network resources) in the urban political arena, or at other governmental levels, to effect urban change in their own interests flatly contradicts any plausible theory of democracy. It may or may not be eradicable, but it is a limit on democracy nonetheless. Given this, although incorporating the private sector in urban management may be justified in terms of managerial flexibility, efficiency or expertise, or more generally as being a means of realizing the popular will or responding to urban citizens' needs, it is not in itself democratic.

It could be argued that in any political division of labour there is a risk that specialized political agents will appropriate public resources, or their positional power more generally, to their own benefit. One of the rationales for involving businesses and other groups in managing cities, after all, is the negative perception of specialized politicians in many contemporary democracies. This brings us to the second problem for democracy associated with the new governance, which is the problem of accountability. The potential ability of citizens to hold specifiable political agents accountable through democratic procedures is central to how democracy works, and specialized politicians act in the context of mechanisms of public accountability, however limited, while businesses and other governance participants do not. In the governance context, where businesses and other organizations play key policy roles, effective mechanisms of democratic accountability are hard to envisage without undermining the very non-hierarchical virtues governance is said to encourage. Jessop (2001) highlights this problem in the context of an argument that governance embodies an alternative form of rule ('heterarchy') from that of the market ('anarchy') or state ('hierarchy'). In the context of organizations linking parts of government, interest groups, businesses, neighbourhood associations and others in horizontal rather than vertical partnership relationships, it can be difficult to know quite who or what is responsible when something goes wrong, or when something succeeds. In these sorts of organizational arrangement, there are always likely to be tensions among the participants, for reasons of mistrust, fear of breach of organizational or commercial confidentiality, different interests in outcomes defined over different time-horizons and different understandings of means and ends. As Jessop suggests, these sorts of tensions are arguably important in making governance effective. In this context, and especially where there is turnover in members of any given arrangement, the clarity of responsibility for actions potentially characteristic of government is absent, partners have strong incentives not to take responsibility if problems arise, and conflict over issues of responsibility are potentially problematic.

If governance arrangements are means to public ends, then the issue of responsibility matters, even where elements of the public are involved among a set of horizontally linked partners. Jessop's arguments take up the idea of governance always requiring 'metagovernance' (the governance of governance), and he and others have linked this to the ongoing responsibilities of states in steering the new arrangements. The degree to which the state can or should plausibly be held responsible for problems in urban policy implementation where responsibility has been devolved to a partnership or similar is nonetheless unclear. Moreover, in this situation of blurry private–public boundaries, complexity really is a problem for democratic publics: in terms of their coping with information fragmentation and overload as differentiated initiatives proliferate; in terms of their capacity to monitor their ongoing work; and because of the necessarily selective public involvement they afford. In his book on the complexities of the new governance in the UK, Rhodes (1997) realizes that the proliferation of these arrangements, combined with complex and overlapping policy networks cutting across jurisdictions and scales, raises severe problems for prevalent understandings of democratic accountability. He does not, however, consider most of the available solutions adequate, suggesting that simple solutions cannot cope with complex problems, and that 'messy problems need messy solutions' (Rhodes, 1997: 21, see also 197–8). One suggestion is that we remodel what we understand accountability to be, thinking of it as policy-domain or policy-network specific. This suggestion relates to questions about the different status of 'the people' in democracy – as citizens or as consumers of public goods and services (Burns, 2000). One thing is certain: even the task of specifying what kinds of 'messy solution' to the accountability problems that new governance arrangements cause as they tackle 'messy problems' is a difficult task for theorists of the phenomenon. This is not comforting from the point of view of squaring governance and democracy in the near future.

The third problem for democracy in the context of the new governance relates to what are arguably its most democratic aspects. Defenders of experimentation with new arrangements for tackling local problems can readily point to the energizing effect that public participation in partnerships, forums, citizen juries and so on can have in making cities work better, and in creating more fruitful relationships between citizens and their governments (see Goss, 2001). Governance, on this view, is a potential antidote to civic privatism, affording opportunities for building social capital and meaningful inclusion in the political process for people on issues that matter to them most. This line of argument can be hard to resist, although the jury is out on the degree to which neighbourhood representation and other modes of citizen incorporation in the new governance are empirically able to decisively modify, let alone determine, the mechanisms and goals of partnership organizations in which state and business organizations are often dominant partners. Unless the mechanisms of democracy are viewed in a narrow way in terms of periodic elections on the one hand, and rights to organize and protest on the other, experimentation with new means of connecting citizens to specific policy initiatives is nonetheless clearly attractive, especially in the light of the accountability problems discussed above.

But differentiated judgement is obviously required. Some fashionable forms of governance, such as Business Improvement Districts, are financed and managed by compulsory levies on local businesses or landowners. They are organized in such a way that, however effective they may be in re-shaping areas of cities in the interests of these contributors, and perhaps thereby in the interests of sections of the urban public at large, it is hard to see them becoming effective democratic vehicles. Similarly, large regeneration projects, with all the financial investment, risk and potential profit that they involve, are situations where it is unlikely that, in general, selective involvement of local citizens can decisively alter or frame their main goals unless the interests of other partners are congruent with public views. Any large-scale participation of the public around city-wide strategic plans, currently a phenomenon common in cities across the world, involves, of necessity, a huge amount of interpretative work by city officials and other actors. This inevitably opens a gap between the raw material of the local forum and the finished scheme, a gap that is just as likely to produce disaffection from the political process as the gap between citizens and their representatives that proponents of governance are keen to close up. In other words, participatory governance is just as likely to breed cynicism and mistrust in urban publics as established forms of delegated responsibility, and those situations where it is less likely to do so are not easy to specify in advance.

The necessarily selective inclusion entailed by more dialogic or participatory governance arrangements is in itself a problem. It is difficult to disengage the value of democracy from values of equality and universality. Whatever 'the people' is, and whatever its geography, democracy seems to demand that its members have an equal potential influence and that this influence is equally shared. In the contemporary differentiated urban polity, with all its multiplying points of access for different groups of citizens, usually selected in terms of specific criteria of residence or group or organization membership, there are a host of issues opened up about the effects that such dispersed and selective involvement has for the value of democratic membership. Combined with the problems of complexity and accountability outlined above, familiar problems of biographical availability and participation fatigue also bedevil attempts to run cities on the basis of shifting special purpose partnership arrangements, heightening the risks of generating an even more stratified hierarchy of effective political citizenship rights than that associated with 'government'.

The 'new governance' in cities, then, raises many troubling questions. How do we make differentiated normative judgements about the roles of different 'sectors' of the urban social fabric in governance? How far is a more differentiated and interactive involvement of urban political actors in urban policy-making compatible with democracy? What constitutes an urban political 'constituency' under these conditions of selective and differentiated participation? In addition, we have to ask how far any of the new arrangements, which inevitably entangle actors and interests operating over a variety of geographic scales, are plausibly 'urban' in any case? As in the first two sections of the chapter, complexity creates difficulties here for any straightforward view of cities as sites of democracy. Nonetheless, the emergence of experimental forms of governance in which the theme of reinventing democracy, as well as reinventing government, is at stake has to be seen as generally positive in democratic terms. It helps to keep open imaginative space for asking questions about the flexibility of what counts as democratic institutions and practices in cities (and of course at other scales). I have already noted Legalès' suggestion that cities are fascinating because they are unavoidably incomplete societies, particular combinations and intersections of actors and relationships with particular political implications. It is obvious from this that different sorts of arrangement in different cities represent potentially equally viable attempts at institutionalizing democracy. Contemporary urban politics involves a plethora of different kinds of activity, procedure, and device. Many debates, within and without the academy, revolve around the competing claims of these various elements of political life to embody democracy or to be more democratizing than others. So, for example, public participation in planning and urban policy-making (Healey, 1997), insurgent activities by social movements (Sandercock, 1998), or elections, can all readily be argued to be the most important foci of aspirations to urban democracy. Different styles of politics, from the symbolic politics of the carnival or festival, through the disruptions of direct action, to the negotiations of deliberative decision-making and the silence of the voting process can be invoked. So can different loci of urban democracy, from the street, via the clubs and organizations often said to build social capital, to City Hall. Different processes, styles and spaces can be more or less marked as pertaining to 'the state' or 'civil society', to the formal or informal dimensions of political life, or to the public or the private. Democracy should be, as Amin and Thrift (2002) suggest, about the politicization of means and procedures as well as of participation as such. Debates about different styles, channels and spaces of democracy are a vital dimension of democracy, although their purpose cannot be to insist that one urban form trumps all the others. For all the difficulties new forms of governance pose in democratic terms, it may be through attempts to puzzle through their implications that a welcome flexibility or openness about how best to translate democracy into urban institutions in different places comes about. The assumption of easily defined urban 'peoples' communicating straightforwardly with their governments does little to encourage this space for transformation and invention. Pluralizing the actors and institutional forms involved in running cities, in other words, should help us develop a more plural, open set of conceptions of what democracy in cities can involve.

Conclusion

This chapter has identified three problems for the way we think about contemporary cities as actual or potential spaces of democracy. Urban societies are complex and differentiated; urban governments are enmeshed in a range of unavoidable political relationships; and 'government' in cities is as fuzzy and disarticulated as are urban popular constituencies. These problems are not obscure ones requiring any complex social theory to make them evident. But they certainly need conceptual innovation and debate for potential solutions to them to be identified. They are at times covered over by an ideologically compelling but limiting spatial imaginary of democracy. This involves a shorthand in which we imagine democracy involving a directed flow of influence, will or pressure from a coherent social body or constituency – 'the people' – to a singular government. The latter acts as an instrument capable of transforming this flow into a stream of popularly mandated policy, thus creating a closed circuit of clear and consequential political communication. Tensions set up by this picture of directed flows between coherently imagined 'peoples' and 'governments' animate many debates about democracy. Democracy seems to demand a transmission process from 'the people' if it is to make any sense, yet this necessary flow sets up the possibility of democratic failure though misinterpretation and distortion.

This picture of democracy implies a simple geography, where the longer the flows of communication between demos and political power, the greater are the possibilities for democracy's corruption and for the substitution of other agendas for those of the people. Specifically, this makes it more likely that various mediating agents (governmental organizations at various scales, political parties, bureaucrats, courts, media institutions) will become necessary parts of the transmission process, and that they will come to have the positional power to re-define the scope, content and outcome of the popular message. In certain versions of democratic theory, the circuit can be shortened and purified through various forms of participatory or direct popular rule that bypass specialized institutions of government and political parties. From the point of view of urban democracy, the flow of influence, desires, needs, pressures, interests or choices (understood as a sort of emanation from the people) is supposed to be easier to read and less riskily transmitted at a local scale. Hence, the special status which cities, local autonomy, and associated 'on the ground' experiments in governance have often had in thinking about democracy's pasts and futures.

This way of conceptualizing the geography of democracy is highly limiting. It can lead to implausible assumptions about the priority of the local, and of cities in particular, as privileged democratic spaces, and to the related assumption that relations between cities and other levels of government are limits on, rather than being necessarily constitutive of, democracy. It can also lead to an overvaluation of new forms of governance on the grounds that they link citizens more directly into policy-making and policy-implementing networks, and they are more likely to express and make effective the popular will than established, more hierarchical arrangements which embody clearer lines of accountability. It is tempting to say, in the face of the realities of contemporary urban life, that democracy is in effect made impossible by the complexity of the spaces with which it has to reconcile itself. The argument of this chapter is that there is no basic blueprint for democracy that is valid for all times and places. Part of what makes democracy work is the sense that it is necessarily about contesting and changing the ways in which citizens communicate with power over different distances, how they oppose it, and how they try to hold it accountable. Democracy is about the re-invention of these relations in changing situations in different geopolitical contexts. Rather than tying it down to specific spaces, whether national or urban, it is better to acknowledge that democracy has only thrived to the extent it has because it has involved the continuous development of practices, institutions and forms of communication that have made it socially and spatially complex.

These conclusions may seem rather negative from the point of view of debates about cities and democracy. But if they are so, it is only in relation to a set of expectations about the affinities between cities and democracy that are unhelpful, and which insufficiently recognize the contingent, open and indeterminate character of democratic practices and spaces. Cities are certainly spaces where democracy should matter, work, be tested and extended, even if they are not in some way spatially or socially fundamental in our thinking about this form of rule. Thinking beyond simple stories about how the microsociological character of cities, about autonomy, or about how the closeness of urban populations to government make real democracy more attainable, should actually make it easier to bridge the gaps between democratic theory and empirically grounded debates about urban programmes and institutions. Paradoxically, then, cities may be important for thinking about democracy because this necessarily involves countering the mythology of the city as a space with special affinities with democratic practices.

References

Amin, A. (1997) 'Beyond associative democracy', *New Political Economy* 1, 3: 309–33.

Amin, A. and Thrift, N. (2002) *Cities: Reimagining the Urban*. Cambridge: Polity Press.

Boden, D. and Molotch, H. (1994) 'The compulsion of proximity', in Friedmann, J. and Molotch, H. (eds), *NowHere: Space, Time and Modernity*. Berkeley: University of California Press.

Borja, J. and Castells, M. (1997) *Local and Global*. London: Earthscan.

Burns, D. (2000) 'Can local democracy survive governance?', *Urban Studies*, 37: 963–74.

Castells, M. (1977) *The Urban Question*. Cambridge, MA: MIT Press.

Clark, G. (1985) *Judges and the City: Interpreting Local Autonomy*. Chicago: University of Chicago Press.

Dahl, R. (1960) *Who Governs? Democracy and Power in an American City*. New Haven, CT: Yale University Press.

Dahl, R. (1989) *Democracy and its Critics*. New Haven, CT: Yale University Press.

Goss, S. (2001) *Making Local Governance Work: Networks, Relationships and the Management of Change*. London: Palgrave.

Hayek, F. (1960) *The Constitution of Liberty*. London: Routledge.

Healey, P. (1997) *Collaborative Planning*. London: Macmillan.

Held, D. (1995) *Democracy and the Global Order: From the Modern State to Cosmopolitan Governance*. Cambridge: Polity Press.

Hurley, S. (1999) 'Rationality, democracy and leaky boundaries: vertical vs. horizontal modularity', in Shapiro, S. and Hacker-Cordón, C. (eds), *Democracy's Edges*. Cambridge: Cambridge University Press.

Isin, E. (ed.) (2000) *Democracy, Citizenship and the Global City*. London: Routledge.

Jenkins, S. (1998a) 'Let's not pick a clapped-out old has-been'. *The London Evening Standard*, 16 January: 5.

Jenkins, S. (1998b) 'Who's pulling the strings?', *The London Evening Standard*, 17 December: 6.

Jessop, B. (2001) 'Governance failure', in Stoker, G. (ed.) *The New Politics of British Local Governance*. London: Macmillan.

John, P. (2001) *Local Governance in Western Europe*. London: Sage.

Judge, D. (1995) 'Pluralism', in Judge, D., Stoker, G. and Wolman, H. (eds), *Theories of Urban Politics*. London: Sage.

Labour Party (UK) (1996) *A Voice for London* (Consultation Document). London: The Labour Party.

Legalès, P. (2002) *European Cities: Social Conflicts and Governance*. Oxford: Oxford University Press.

Low, M. (1997) 'Representation unbound: globalisation and democracy', in Cox, K. (ed.), *Spaces of Democracy*. New York: Guilford.

Manin, B. (1997) *The Principles of Representative Government*. Cambridge: Cambridge University Press.

Pecorella F. (1987) 'Fiscal crises and regime change in New York'. In Stone C. and Sanders, H. (eds), *The Politics of Urban Development*. Lawrence: University of Kansas Press.

Peterson, P. (1981) *City Limits*. Chicago: University of Chicago Press.

Phillips, T. (1997) 'The real revolution is in London', *The Independent*, 26 July: 17.

Polsby, N. (1980) *Community Power and Political Theory*. (Second Edition). New Haven, CT: Yale University Press.

Putnam, R. (1993) *Making Democracy Work: Civic Traditions in Modern Italy*. Cambridge, MA: Harvard University Press.

Rhodes, R. (1997) *Understanding Governance*. Milton Keynes: Open University Press.

Rodriguez-Pose, A. and Gill, N. (2003) 'The global trend towards devolution and its implications', *Environment and Planning C: Government and Policy*, 21: 333–51.

Sandercock, L. (1998) *Towards Cosmopolis*. London: Wiley.

Scott, A. (1988) *Metropolis*. Berkeley: University of California Press.

Sennett, R. (1977) *The Fall of Public Man*. London: Faber.

Shapiro, I. and Hacker-Cordón, C. (eds) (1999) *Democracy's Edges*. Cambridge: Cambridge University Press.

Shefter, M. (1996) 'Political incorporation and political extrusion: party politics and social forces in postwar New York', in *Political Parties and the State*. Princeton, NJ: Princeton University Press.

Stone, C. (1993) 'Urban regimes and the capacity to govern: a political economy approach', *Journal of Urban Affairs*, 15: 1–28.

Taylor, P. (1995) 'World cities and territorial states: the rise and fall of their mutuality', in Knox, P. and Taylor, P. (eds), *World Cities in a World System*. Cambridge: Cambridge University Press.

UK Government (2001) *Modern Local Government: In Touch with the People*. London: HMSO.

Wolman, H. (1995) 'Autonomy and city limits', in Judge, D., Stoker, G. and Wolman, H. (eds), *Theories of Urban Politics*. London: Sage.

Young, I. M. (1990) *Justice and the Politics of Difference*. Princeton, NJ: Princeton University Press.

Zolo, D. (1990) *Democracy and Complexity*. Cambridge: Polity Press.

Zukin, S. (1995) *The Cultures of Cities*. Oxford: Blackwell.

8   Spaces of Public and Private: Locating Politics

Lynn A. Staeheli and Don Mitchell

The spaces of democracy are necessarily plural. They include, most obviously, the spaces constructed in and through the state, such as the nation-state, cities, governments, and public spaces. But we should also consider the possibilities for democracy in other spaces that may not seem so obviously political or so obviously constructed and maintained by the state. These spaces may be associated with the home, neighborhood, commerce, work, worship, or any gathering place. We argue in this chapter that these are not just spaces of democracy or of politics; they are spaces defined by unequal and differential access. As such, they offer differing possibilities for politics that are shaped as much by the agents that operate within those spaces as by the spaces themselves.

Central to our argument is that the political opportunities associated with various places are often assessed in terms of publicity and privacy. A host of political theorists going back to Aristotle and continuing to the liberal theorists of today have argued that politics and democracy take place in the public realm (Weintraub, 1995). But while locating politics in the public sphere, these theorists also recognise the importance of the private sphere. In the abstract, the private sphere is one of negative freedoms (in other words, a sphere free from interference) in which political ideas can be constructed and autonomous citizens are nurtured (Cohen, 1995; Elshtain, 1995). We argue that the play between public and private and the attempts to redefine the boundaries between these are critical to the potential for democratic action and politics.

Our analysis begins with what may seem to be a counter-intuitive argument about the erosion of privacy. It is, after all, not the obvious starting point if we are correct that most theorists locate politics in public. We argue, however, that the public and private are imbricated, and therefore the erosion of privacy causes a rearrangement of the spaces of democracy in ways that implicate dimensions of social difference. We use this as a way to demonstrate the different meanings of public and private that are invoked in debates about the changing nature of democracy. While the wide range of meanings associated with the terms public and private is often derided and lead to calls to abandon the terms, we argue that these different meanings of publicity and privacy, and their concomitant spaces, allow different sorts of politics and political opportunities. The second section of the chapter considers the implications of the diffusion of democracy for our understandings of the politics of public and private. It must be recognised that most theories of democracy and of publicity and privacy are rooted in Western thought and experience, so we use this section to explore the implications of the diffusion of Western institutions under the label of democratisation (Bell and Staeheli, 2001) for publicity and privacy. We conclude the chapter with a consideration of the implications of these arguments for what it means to engage in a struggle for spaces of democracy and justice.

Public, Private and the Location of Politics

While the public sphere has been theorised as the space of politics, it is a space that is also constituted in and through privacy.[1] The private sphere is seen as a space of 'negative freedom' or of freedom from governmental or societal intrusion. This is a space where opinions may be expressed and actions taken without fear of state surveillance or sanction; as such, the private sphere is held to be a place where civil liberties protect individuals in the creation of autonomous political subjects (Elshtain, 1995). The private sphere is also supposed to be the space in which ideas can be developed and decisions made in relative safety from social pressures. It is a forum for discussion in which ideas can be debated without regard to cannons of civility, cannons that may limit either the types of issues raised or the ways in which issues are resolved. The private sphere does not privilege particular forms of discourse, and, as such, may actually allow a wider range of voices and positions to be articulated. Historically, this has been important for women and for other groups that may not have been seen as capable of the rationality required for deliberation in public (Marston, 1990; Ryan, 1990; Skocpol, 1992).

While the private sphere is seen as important for democracy and politics, there are several contemporary challenges that are both theoretical and practical. Many feminists have argued that the private sphere is a place of patriarchal privilege in which men dominate women. The protection from state interference implied with privacy, for example, has been seen as an impediment to laws that protect women and children from domestic violence (Schechter, 1992); if the private space of the home is to be inviolate, then law enforcement should not intrude on what is constructed as a domestic, rather than public, matter. New technologies that track activities (from credit card transactions to websites visited to preferences for various branded consumer products) previously understood as private are now sold and shared among business and governmental agents. In the United States, new rules and legislation enacted after the tragedies of September 11, 2001 have allowed government access to conversations, such as those between clients and attorneys, that were previously held to be private. US Attorney General John Ashcroft argued in the latter instance that there was a public interest in knowing what some accused prisoners might say to their lawyers (Purdy, 2001). At the same time, and perhaps paradoxically, the adoption of public nuisance laws in cities, coupled with the privatisation of public spaces, transforms and limits the public nature of city spaces (Zukin, 1995; Mitchell, 1997; Smith, 1998).

Jean Bethke Elshtain (1995), Jean Cohen (1995) and Judith Squires (1994) argue that these trends have reconfigured the relations between public and private in ways that may not be intended. Elshtain, in particular, is critical of what she calls a 'displacement of politics' in which the erosion of privacy and the blurring between what remains of privacy and the public sphere is nearly complete. She argues (Elshtain, 1995: 170) that '[i]f all conceptual boundaries are blurred and all distinctions between public and private are eliminated, then by definition, no politics can exist'. She continues that the feminist mantra 'the personal is political' and other identity-based slogans downplay the different logics that structure public action and private intimacy. Squires (1994) argues that one reason that public and private have become fused in some analyses has to do with the complicated and often divergent meanings of privacy. Privacy can imply deprivation (Arendt, 1958), confidentiality, autonomy, or intimacy, and the political implications of each of these are quite different. Squires argues that privacy is most frequently invoked (in public discourse and legal opinions) as both a sphere of negative freedom and as domestic and intimate to the point that these two meanings are often conflated. Yet the understanding of the private as a sphere of civil liberties and negative freedom is liberal, whereas the meaning of private as domestic is often patriarchal. As such, the two readings are mixed, and indeed, contradictory. The issue is further compounded when property relations and the privatisation of urban public space are considered. This privatisation portends what Jeremy Waldron (1991) describes, sarcastically, as a 'libertarian paradise'. This paradise is one in which *all* property, and hence in capitalist societies all *space,* is privately owned, and the specific negative freedom of each owner is protected. In such a paradise, not only would there be no (material) public sphere at all, since one could only meet with others 'in public' if she or he were specifically invited to do so by the owner of the property, but those who did not have access to property (as either renters or owners) simply could not *be –* at least without breaking trespass laws. Theorists of the public sphere and its politics thus need to pay attention to relations of *property,* and their very real limitations, as they seek to assess the possibilities for an inclusive public (Blomley, 1997; 2004).

In fact, the meanings highlighted above are not exhaustive of the ways in which privacy is defined, and, as such, they are only indicative of the complexity of the term. And since publicity and privacy are defined in relation to each other, the relationship between the two spheres is multivalent and may shift over time and across different settings for different social groups (see Benhabib, 1992, 1996). Weintraub (1995) identifies at least four models through which public and private are defined in political theory. These include a liberal-economistic model in which publicity is associated with the state and privacy is associated with the market; a republican-virtue model that posits the public sphere as community and citisenship, which are held to be distinct from the state and the market; a model of sociability in which publicity is defined in terms of self-representation, display and performance; and a feminist model in which the public refers to the state and economy while the private refers to the domestic and familial. These models rely on different theoretical constructs to identify public and private, but often in ways that are incompatible with each other (e.g., the state and market are constructed as private in the republican-virtue model and as public in the feminist model). To these definitions, geographers and legal scholars add an additional definition that considers the role of property. These definitions differ from the theoretical definitions discussed by Weintraub in that they refer to material spaces, rather than spaces in the abstract. The significance of this distinction will become clear in the following discussion.

Inevitably, there is confusion when authors write of public and private without defining their terms. In addition, there are gaps in the models that leave important arenas of action out of the conceptualisation of either public or private. From the perspective of democracy and politics, one of the most important arenas overlooked in most models relates to community. This has led some analysts to propose a third arena to their model, as a way of recognising an important sphere that is ambiguous with respect to publicity and privacy (Moore Milroy and Wismer, 1994). Others argue that the mistake is in looking for *one* public, when there are, in fact, multiple publics. Nancy Fraser (1990) is probably best known in this regard, arguing that the idea of a single public sphere privileges and universalises the partial perspectives of a group of elite, white males. She proposes that we think of multiple publics in which members of marginalised groups can articulate interests and strategies, develop political positions, and from which groups can speak to other publics. Chantal Mouffe (1992) and Bruce Robbins (1993) argue that such a conceptualisation of the public sphere could involve recognition that each political agent is positioned with respect to multiple publics, which both complicates and enriches their political ideas. Against these arguments for building a more complex understanding of publicity are those who argue that the idea of public and private should be discarded as meaningless, confusing, and as reflective of the problems of dichotomous thinking (Rose, 1993; Gibson-Graham, 1996).

Yet there is something compelling about the idea of publicity and privacy that is enduring. Barrington Moore (1984) and Brian Fay (1975) have argued that every society makes some distinction between public and private and that this distinction is important in lending coherence to social life. Robbins (1993) has argued that it provides an ideal and a foil against interests that promote private greed at the expense of the general public. Elshtain (1995) agrees, noting that the idea of the public focuses attention on what is shared or common in a polity characterised by injustice and subordination of marginalised groups. This shared understanding is a necessary precondition for a 'practical politics' that is built on temporary alliances or coalitions between representatives of diverse groups as they attempt to address concrete concerns. This, she argues, can only happen if we recognise the tensions that exist between spheres – tensions that create a space for politics:

> [T]he public world itself must nurture and sustain a set of ethical imperatives that include a commitment to preserve, protect, and defend human beings in their capacities as private persons, *as well as* to encourage and enable men and women alike to partake in the practical activity of politics. Such an ideal seeks to keep alive rather than to eliminate tensions between diverse spheres and competing values and purposes [...]. Only in the space opened up by the ongoing choreography of [public and private] can politics exist – or at least any politics that deserves to be called democratic. (Elshtain, 1995: 180)

If, as Elshtain suggests, the tension between public and private creates a space for politics, then two questions arise: What kinds of politics are possible? And in what sort of spaces do these politics develop?

Above all, politics is messy. Michael Brown (1999) argues that every political issue can be characterised in terms of multiple senses or dimensions of public and private. As such, political agents can gain new insights into issues if they recognise the multidimensionality of publicity and privacy, and this, in turn, can open new strategies for action. The ability to move across these dimensions of public and private are also an important, if under-examined, element of the politics of mobility and access, of the 'power-geometry' described by Doreen Massey (1994) and Jennifer Hyndman (2004). Similarly, Tim Cresswell (1996) has demonstrated the significance of taking a problem outside the realm or space with which it is associated and the potential for political action that is 'out of place'. In one example, he notes that women were startlingly effective (though also quite controversial) when they engaged in protests over nuclear proliferation. By protesting in public, not only did the women at Greenham Common in the United Kingdom transgress the spaces to which they were assumed to belong, they did so by bringing different senses of public and private to bear in their protests. Such action does two things: it highlights the very different sorts of public interest embedded in any issue, and it opens space for action that might not have been conceivable within a singular, fixed understanding of public and private. As such, it is a paradoxical *space* of politics (Rose, 1993) that presents the opportunity for creating new ideas, understandings and strategies for action.

But what is the *structure* of this 'paradoxical' space? As Howell (1993) and others have argued, much public sphere theory, especially as it has developed in relation to Habermas's (1989) conceptualisation, is aspatial. Or to put that more accurately, the spatial structure of the public sphere is under-theorised. A fuller theorisation of the spatiality of the public sphere must begin, as we have noted, from the understanding that public and private *spaces* should not be conflated with public and private *actions* (Staeheli, 1996). But such an understanding makes more imperative the need to clearly understand the material construction of space and the relations of power that give it form. Consider, as an example, the workplace. Wherever political theorists may 'place' the office or factory floor, either as part of the public or private sphere, the fact remains that it is a space defined by law and social practice that significantly limits the sorts of action that may occur there, and it limits them in a way that they may not be limited outside the factory gates or beyond the employee parking lot. Many workplaces are seemingly open in that ideas are debated, and, as noted, many theorists locate the economy in the public sphere. But against these practices and theories are legal rulings that construct the workplace as a space capable of tight control and in which political activities may be limited.

As more and more 'public' functions have moved to the privately owned spaces of the shopping center and corporate plaza, in the United States, the courts have refused to extend the protections of public spaces. Increasingly, property law makes a difference in setting the conditions for public activity. This becomes even more significant if one considers the ways in which private spaces may be shaped by multiple meanings of privacy. Recall Squires' (1994) argument that the home may be simultaneously a private space of liberal freedom and of patriarchal oppression. Now add to the mix the changes in the home as a workplace. While the home remains a site of domestic labor, it is often also a site of paid labor over which employers (from outside the home) may wish to exert control. Telecommuting, extended working hours, and the common practice of bringing work to the home means that employers may influence and may monitor activities in a private space that is not one over which they have property rights. Thus, the home, always paradoxical, is changing as a space of privacy. It may remain 'private', but the agents who have an interest in that space may not have legal ownership over the space. And in a cruel twist, the multiple senses in which spaces are constructed as private can lead to the erosion of privacy in the liberal sense, as the spaces may become subject to greater surveillance and as they are mediated by outside agents.

These examples suggest that it is inadequate to argue that certain actions 'open a space for politics' unless we pay very careful attention to the material constitution of that space: how it is shaped, who has access to it, what sorts of laws govern it, and how the actions that develop in it are spread across geographical networks of communication. The combined influence of the privatisation of public space and the retrenchment of liberal freedoms associated with private spaces suggests there is less and less space for democratic politics. This is a particular concern to the extent that the trends result in exclusions, either of voices or of people. Such exclusions seem inevitable, and are already evident. In addition, the potential for transgression, that is, for taking actions out of their 'appropriate' place, is transformed, and not necessarily in ways that enhance democratic practice. As Waldron (1991) suggests, there may simply not be spaces for politics, resulting not just in a *displacement* of politics (which is Elshtain's concern), but the *impossibility* of politics. Finally, the importance of legal doctrine in shaping the material spaces of public and private means that any political theory which invokes ideas of 'public' needs to specify which countries or actually existing democracies (Fraser, 1990) to which their theories apply. While this is true generally, it may be particularly important for countries struggling to build democratic institutions and, in some cases, to conceptualise and enact alternative forms of democratic practice.

Publicity, Privacy and the Diffusion of Democracy

One of the most heralded trends during the last quarter of the twentieth century was the so-called 'third wave of democratisation'. Samuel Huntington (1991) and others cheered the establishment of democratic institutions in over 100 countries during that period. The trend of the diffusion of democratic institutions *is* important, as it is associated with concrete advances involving open, competitive elections, a broadening of rights to political participation, and new protections for civil liberties in a growing number of countries (O'Loughlin et al., 1998). Without belittling the significance of these trends, however, it should be recognised that democratic institutions were installed in many places through decidedly undemocratic means, forced through military action, external pressures, and/or the actions of domestic elites (Gill, 2000) with little regard (indeed, sometimes with intentional disregard) for established social and political practices within a country. While there are many implications of this imposition, we focus here on the nature of publicity and privacy.

Put simply, many of the arguments about the importance of the public sphere (or about civil society, its analog in some literature) assume that democratic institutions work because of the functioning of a well-developed, robust public sphere. The public sphere and civil society, in a sense, are seen as precursors or requisites for the effective operation of institutions of governance (e.g., Habermas, 1989; Putnam, 1993). But as Karl (1990) and others have argued, there is no guarantee that the imposition of institutions will be successful in places with poorly developed senses of public or in places where notions of publicity differ from those of the West. Three short examples indicate the problems, and also the potential, when Western institutions and ideals regarding publicity may not match with long-established practices and traditions. Each of these examples demonstrates the importance of politics along the borders between publicity and privacy in defining what can be addressed through politics and in shaping access to politics. And as so often happens, these challenges are highlighted as democracy is diffused to other places and reflect back on the assumptions and practices that define democracy in the West. As such, they provide an opportunity to reflect critically on the roles of publicity and privacy in shaping the spaces of democracy.

The first example comes from the former Soviet Union. Barbara Einhorn (1993) argues that dissident movements were launched in homes during the Soviet regime because these spaces were relatively free from state observation and control. At some level, this role for private spaces is entirely consistent with ideas of privacy that argue this is the realm where autonomous (but not necessarily disembodied) political subjects can develop. Yet there was more that happened in the private spaces of the home than just the nurturing of political subjects. Issues were debated; strategies and plans were developed. In many ways, these spaces constituted a public sphere in which a nascent civil society was developed. It was, of course, exclusionary, as participants sought protection from an intrusive state. But the public sphere that developed was probably no more exclusionary than the public sphere of Athenian democracy or eighteenth-century coffee houses (see Habermas, 1989). Instead, the challenge posed by this example is that democratic potential was enhanced by privacy. A defining characteristic of democratic institutions and practices is publicity in the sense of visibility (Hartley, 1992; Goheen, 1998),[2] as it is only through visibility that legitimacy and accountability can be enforced. Yet, in this case, the spaces that best fostered debate and action to ensure publicity were themselves private.

It is, of course, possible to point to many examples within the West and within established democracies of this kind of politics enacted in private. Staeheli (1996), for example, has argued that publicity should not simply be read from the spaces in which activity occurs. It may not, therefore, seem obvious why the above example poses a challenge for the diffusion of democracy. The significance, however, lies in the way in which politics is perceived and the implications that hold for the spaces of democracy. Einhorn (1993) argues that 'politics' became associated with an authoritarian state and, thereby, was itself something to be resisted. Indeed, the private spaces of the home were understood as spaces of 'anti-politics', even as they were sometimes sites of resistance movements. The challenge, then, is not that political discussion happened in private. It is that the public sphere was, and still may be, a place where government, not democracy, ruled. This, in turn, has important implications for the ability and willingness of a citizenry to participate in politics that are more public, such as those that would force accountability and legitimacy on a government, or even to reach agreement as to what those terms might mean. It is difficult to see how any politics which aims to make government responsive to the public can be achieved within private, material spaces or without entering the public realm.

The second example also speaks to the nature of public spaces and the ways they may or may not facilitate discussion of the common good. The idea of the common good or common understanding is a bedrock of democratic community and is advocated by theorists who take quite different political positions (e.g., compare Baechler, 1993; Etzioni, 1993; Robbins, 1993; and Elshtain, 1995).

Jon Goss (2000), however, argues that in many parts of the developing world, formal public spaces were essentially colonial impositions that reflected two conflicting ideologies. The first was a desire to imprint on the landscape a separation of public and private by building a space for civic purposes. This separation, he argues, is thoroughly Western. The second ideology was exclusionary, as the public spaces of the plaza were regulated in terms of function. By regulating function, the colonial state effectively regulated access to the space on the part of colonised peoples (see also Wilson, 1997). Goss argues that squatters in the Philippines resisted the logic of both ideologies. They attempted to create a space that was multifunctional (rather than just civic) and that was multi-class and multi-ethnic (rather than exclusionary). In this public space, politics were not hived away from commerce, from leisure, or even from living (Abbas, 1997; Law, 2002). This represents a rather different conceptualisation of publicity than that embedded in much Western political theory, as participants in the squatters movement seemed to promote a seamlessness between, or perhaps a jumbling together of, public and private that many Western theories reject. Yet if the struggles over space are indicative of the social norms held by the squatters, it seems important to account for this in our theories of democracy, or at least to not reject them out of hand as anti-democratic. Indeed, rather than the 'displacement of politics' feared by Elshtain, the struggle over inclusion and how it can be accomplished is central to building and consolidating democracy in postcolonial settings and elsewhere.

The final example highlights the difficulties when formally democratic institutions are overlain on cultures and customs that do not recognise the personhood of all citizens or inhabitants. When this happens, the private sphere can become a site of domination, oppression, and even terror. Richa Nagar (2000) recounts the work of women's organisations in northern India that are trying to eradicate domestic violence. India is widely proclaimed as the world's largest democracy, and democratic institutions and the formal protection of civil liberties have been in place for decades. But domestic violence and the murder of wives continue. In response to its persistence, women's organisations engage in street theater to expose the problem and to put pressure on those who tolerate violence. In itself, this is a practice that relies on a spatial tactic to bring into public something that happened in private. By bringing this to attention and forcing discussion of it, participants in the street theater rely on members of the public to enforce behaviors that respect the personhood and liberties of women in the private spaces of the home. This example demonstrates, first, that formal institutions that are supposed to guarantee basic civil liberties such as life are ineffective in the face of long-standing social practices that deny the personhood of women. The private space of the home, which is often conceptualised a site of protection or safety in political theory and in Einhorn's example, can also be a site of oppression and terror. Second, the women's organisations were able to move across public and private, to bend and reshape these boundaries, in ways that opened a space for politics that appears to be more effective than operating within the institutional spaces defined by the Indian state and by local acceptance of 'private' behaviors.

Again, this example could easily be recast in terms of the mismatch between institutions and practices in any country, including the United States. As noted, the controversy over domestic violence legislation has often centered on the question of the nature of public interest in behaviors that occur in private spaces, such as the home. Rather than dismiss this example and the others as simple inevitability, it is useful to think about what they might teach us about the way to think about publicity, privacy, and social justice in a world that is more complex than that defined by Western practice and political theory, and in a world in which the reconfiguration of public and private seems on the verge of dislocating politics entirely.

Conclusion: Spaces of Justice and Inclusion

Democracy and justice require some form of privacy that protects people from excessive control or intervention from the state or from third parties (Cohen, 1995). This protection is essential for the development of personhood and of autonomous citizens. At the same time, democracy and justice also depend on some form of visibility and collectivity (Hartley, 1992; Weintraub, 1995). The necessity of both privacy and publicity create tensions and exclusions that can undermine democracy and justice. So how can we think about publicity and privacy in ways that *reinforce* democracy and justice? And how can we do this in a way that recognises the importance of the material constructions of space as those constructions shape publicity and privacy?

Young (2000) and Cohen (1995) argue that the key is to think about the ways in which certain spaces foster or enable inclusion. This requires that we consider the broad forces that shape societies and spaces, not simply think about space in the abstract. As the preceding discussion should have made clear, the forces that structure spaces are multivalent, and they take on different implications for different people at different times and with respect to different issues. As all these 'differences' indicate, it is difficult to imagine a universal and fixed solution.

But where to look for *any* type of solution? If Cohen (1995) is correct that boundaries are necessarily associated with exclusion and control, then it seems reasonable to look to boundaries. We have previously argued that the 'big' trend is toward increasing control over spaces due to the privatisation of public space and the incursion of control by the state and capital over private and even domestic spaces. In a sense, we have argued that powerful institutions have blurred the boundaries between public and private, but in ways that may enhance their own power rather than in ways that enhance democracy and justice. This trend highlights the circular relationship between democracy, justice, and space. As Young (2000: 35) argues, a democracy will only promote justice if there is already a just society. The presence of democratic institutions and practices in societies characterised by structural inequalities will likely reinforce injustice or even extend inequality. To Young's argument, of course, we add the role of space. The presence of democratic institutions and practices in societies in which the spaces of publicity and privacy are characterised by structural inequality are unlikely to allow the development of a citizenry that can use the institutions of democracy to create a more just society. Rather, institutions may be used to reinforce and extend injustice and exclusion.

Against this 'big' picture, we note the smaller transgressions that are intended to expose exclusions and to challenge the structures that create them. Several examples of these transgressions have been raised in this chapter, and each of the actors involved in these examples has incorporated the material construction of spaces in their tactics. The critical point, however, is that they have not used spatial tactics in the same way; their attempts to create more just and inclusive societies have used the *particular* constructions of space as they affect *particular* political agents and problems. Obviously, none of these movements or actions has ended oppression and injustice; rather, these actions represent interventions at the boundaries between public and private that are attempts to reclaim, or even to make, a space for more democratic politics (Cooper, 1998). Rather than being a dislocation of politics, as feared by Elshtain (1995), or an impossibility for politics, as feared by Waldron (1991), these are attempts to make a location for politics, justice and inclusion.

Clearly, the spaces and politics that may be created through these struggles are not fixed. Nor are they necessarily struggles to create a more inclusive, just polity. Rather, these spaces and politics are themselves the objects of ongoing response and struggle. As a result, there is no final geography of publicity, privacy and democracy in either a normative or a material sense. The struggle for democracy, then, may in reality be an ongoing series of transgressions, large and small, which make new, but temporary, locations for politics.

Notes

The research used in this chapter was supported by a grant from the National Science Foundation.
1   This construction or way of describing public and private spheres is one that conflates the type of action with the spaces of action, something criticized by Staeheli (1996). While aware of the difficulties posed by this conflation, we have chosen at this point to use the language of the theorists we discuss.
2   This is also a different usage of the term 'publicity' than has been presented earlier in this chapter, and it is not included in the review from Weintraub (1995). It refers specifically to institutions and practices, rather than the more abstract and nebulous public sphere.

References

Abbas, A. (1997) 'Hong Kong: other histories, other politics', *Public Culture*, 9: 293–313.

Arendt, H. (1958) *The Human Condition*. Chicago: University of Chicago Press.

Baechler, J. (1993) 'Individual, group, and democracy', in J. W. Chapman and I. Shapiro (eds), *Democratic Community*. New York: New York University Press, pp. 15–40.

Bell, J. and Staeheli, L. A. (2001) 'Discourses of diffusion and democratisation', *Political Geography*, 20: 175–195.

Benhabib, S. (1992) 'Models of public space: Hannah Arendt, the liberal tradition, and Jürgen Habermas's, in C. Calhoun (ed.), *Habermas and the Public Sphere*. Cambridge, MA: MIT Press, pp. 73–99.

Benhabib, S. (1996) *Democracy and Difference: Contesting Boundaries of the Political*. Princeton, NJ: Princeton University Press.

Blomley, N. (1997) 'Landscapes of property', *Law and Society Review*, 32: 567–612.

Blomley, N. (2004) *Unsettling the City: Urban Land and the Politics of Property*. New York: Routledge.

Brown, M. (1999) 'Reconceptualising public and private in urban regime theory: governance in AIDS politics', *International Journal of Urban and Regional Research*, 23: 70–87.

Cohen, J. (1995) 'Rethinking privacy: the abortion controversy', in J. Weintraub and K. Kumar (eds), *Public and Private in Thought and Practice: Perspectives on a Grand Dichotomy*. Chicago: University of Chicago Press, pp. 133–165.

Cooper, D. (1998) *Governing Out of Order: Space, Law and the Politics of Belonging*. London: Rivers Oram Press.

Cresswell, T. (1996) *In Place/Out of Place*. Minneapolis, MN: University of Minnesota Press.

Einhorn, B. (1993) *Cinderella Goes to Market: Citizenship, Gender, and Women's Movements in East Central Europe*. London: Verso.

Elshtain, J. B. (1995) 'The displacement of politics', in J. Weintraub and K. Kumar (eds), *Public and Private in Thought and Practice: Perspectives on a Grand Dichotomy*. Chicago: University of Chicago Press, pp. 166–181.

Etzioni, A. (1993) *The Spirit of Community: The Reinvention of American Society*. New York: Touchstone Books.

Fay, B. (1975) *Social Theory and Political Practice*. London: Allen and Unwin.

Fraser, N. (1990) 'Rethinking the public sphere: a contribution to the critique of actually existing democracy', *Social Text*, 25/26: 56–80.

Gibson-Graham, J. K. (1996) *The End of Capitalism (As We Knew It)*. Oxford: Blackwell.

Gill, G. (2000) *The Dynamics of Democratisation: Elites, Civil Society and the Transition Process*. London: Macmillan.

Goheen, P. (1998) 'Public space and the geography of the modern city', *Progress in Human Geography*, 22: 479–496.

Goss, Jon (2000) Interview, 20 July.

Habermas, J. (1989) *The Structural Transformation of the Public Sphere*, translated by T. Burger. Cambridge, MA: MIT Press.

Hartley, J. (1992) *The Politics of Pictures: The Creation of the Public in the Age of Popular Media*. New York: Routledge.

Howell, P. (1993) 'Public space and the public sphere: political theory and the historical geography of modernity,' *Environment and Planning D: Society and Space*, 11: 303–322.

Huntington, S. (1991) *The Third Wave: Democratisation in the Late Twentieth Century*. Norman, OK: University of Oklahoma Press.

Hyndman, J. (2004) 'The (geo)politics of gendered mobility', in L. Staeheli, E. Kofman and L. Peake (eds), *Mapping Gender, Mapping Politics: Feminist Political Geography*. New York: Routledge.

Karl, T. L. (1990) 'Dilemmas of democratisation in Latin America', *Comparative Politics*, 23: 1–21.

Law, L. (2002) 'Defying disappearance: cosmopolitan public spaces in Hong Kong', *Urban Studies*, 39: 1625–1645.

Marston, S. (1990) 'Who are "the people"? Gender, citizenship, and the making of the American nation', *Environment and Planning D: Society and Space*, 8: 449–458.

Massey, D. (1994) *Gender, Space and Politics*. Minneapolis, MN: University of Minnesota Press.

Mitchell, D. (1997) 'The annihilation of space by law: the roots and implications of anti-homeless laws in the United States', *Antipode*, 29: 303–335.

Moore, B. (1984) *Privacy: Studies in Social and Cultural History*. Armonk, NY: M. E. Sharpe.

Moore Milroy, B. and Wismer, S. (1994) 'Communities, work and public/private sphere models', *Gender, Place and Culture*, 1: 71–90.

Mouffe, C. (1992) 'Democratic citizenship and the political community', in C. Mouffe (ed.), *Dimensions of Radical Democracy: Pluralism, Citizenship, Community*. London: Verso, pp. 225–239.

Nagar, R. (2000) '*Mujhe Jawab Do!* (Answer me!): women's grass-roots activism and social spaces in Chitrakoot (India)', *Gender, Place and Culture*, 7: 341–362.

O'Loughlin, J., Ward, M. D., Lofdahl, C. L., Cohen, J. S., Brown, D. S., Reilly, D., Gleditsch, K. S. and Shin, M. (1998) 'The diffusion of democracy, 1946–1994', *Annals of the Association of American Geographers*, 88: 545–574.

Purdy, M. (2001) 'Bush's new rules to fight terror transform the legal landscape', *New York Times*, 25 November, A1, B2.

Putnam, R. (1993) *Making Democracy Work*. Princeton, NJ: Princeton University Press.

Robbins, B. (1993) 'Introduction: the public as phantom', in B. Robbins (ed.), *The Phantom Public Sphere*. Minneapolis, MN: University of Minnesota Press, pp. vii–xxvi.

Rose, G. (1993) *Feminist Geographies*. Minneapolis, MN: University of Minnesota Press.

Ryan, M. (1990) *Women in Public: Between Banners and Ballots, 1825–1880*. Baltimore, MD: Johns Hopkins University Press.

Schechter, S. (1992) *Women and Male Violence: The Visions and Struggles of the Battered Women's Movement*. Boston, MA: South End Press.

Skocpol, T. (1992) *Protecting Soldiers and Mothers: The Political Origins of Social Policy in the United States*. Cambridge, MA: Belknap Press.

Smith, N. (1998) 'Giuliani Time', *Social Text*, 57: 1–20.

Squires, J. (1994) 'Private lives, secluded places: privacy as political possibility', *Environment and Planning D: Society and Space*, 12: 387–401.

Staeheli, L. A. (1996) 'Publicity, privacy, and women's political action', *Environment and Planning D: Society and Space*, 14: 601–619.

Waldron, J. (1991) 'Homelessness and the issue of freedom', *UCLA Law Review*, 39: 295–324.

Weintraub, J. (1995) 'The theory and politics of the public/private distinction', in J. Weintraub and K. Kumar (eds), *Public and Private in Thought and Practice: Perspectives on a Grand Dichotomy*. Chicago: University of Chicago Press, pp. 1–42.

Wilson, C. (1997) *The Myth of Santa Fe: Creating a Modern Regional Tradition*. Albuquerque, NM: University of New Mexico Press.

Young, I. M. (2000) *Inclusion and Democracy*. Oxford: Oxford University Press.

Zukin, S. (1995) *The Cultures of Cities*. Oxford: Blackwell.

9 The Geopolitics of Democracy and Citizenship in Latin America

Gareth A. Jones

This chapter provides an optimistic reading of democracy and citizenship in Latin America. It starts, however, from what might appear to be a set of highly pessimistic accounts of democratization and citizenship, which can be set beside a set of equally forbidding accounts of the social and spatial fragmentation of Latin American cities. These accounts show that the cross currents of opinion on democracy and citizenship in Latin America are in flux and uncertainty. I want to argue, however, that democracy and citizenship present complex geopolitics that need to be understood against more than the 'goodness of fit' of a universal teleology suggested by the neo-institutionalists and transitologists (Massey, 1999). Rather, I draw upon writing on the politics of identity to argue that if we accept that identities are political, constantly unstable and contested, and are also grounded in particular spatial arrangements that, in turn, we understand to be socially produced, then there must be multiple spaces for politics beyond those of an institutional perspective. A vital question, then, becomes how identities are communicated to others in ways that extend democratic potential (Young, 1996). Taking the idea of multiple spatialities seriously, I want to suggest that the formation of spatial representations acts as strategic sites for deepening communicative democracy and the contest of citizenship. Specifically, I propose that one means for making identities public is through the construction of spatial discourses and imaginations that speak to wider audiences, a process which serves to represent certain spaces as public spaces. Spaces become public when they mediate or inform in some way the discourses of the groups that physically occupy them or symbolically invest them with meaning. Public space therefore is performative, where identities are exposed and communicated, interpreted, understood, and transformed. The shortcomings of democracy and citizenship in institutional terms does not undermine, therefore, the case for how a geopolitics can reveal multiple spaces with democratic potential in which everything is possible.

The Rise of Democracy and the Demise of Citizenship

Asked what was the most important thing to have happened in the twentieth century, Amartya Sen concluded that it was the 'rise of democracy'. In justifying his choice, Sen argued that 'the idea of democracy as a universal commitment is quite new, and it is quintessentially a product of the twentieth century' (1999: 4). Elaborating on his response, Sen noted a fundamental change in attitudes towards democracy, from a position in which certain peoples were thought incapable of governing themselves, and were therefore not fit *for* democracy, to one whereby societies were expected to become fit *through* democracy. Sen's reading of the history of democracy is broadly compatible with the 'transitology' studies that have dominated the analysis of political change in Latin America for the past two decades. According to this approach, democracy in the region is increasingly consolidated through a growing respect for the *institutions* of democracy, notably increasingly competitive elections with plural slates, limited interference from the military and a respect for an exchange of mandates (*alternancia*) (Grindle, 2000; Mainwaring and Scully, 1995). The approach provides empirical evidence to support former US Secretary of State George Shultz's congratulatory comment that there are 'more people voting in more elections in more countries than ever before in the history of this hemisphere'.

Mexico represents a good example of the democracy's rise. Characterized by persistent electoral fraud, limited media autonomy, the occasional use of repression against opposition groups and a built-in corporatism and use of public resources that favoured only one political party, by the 1990s Mexico could claim the democratization of its constitutional democratic status. Three weeks after the 1994 national elections, the Director of the Federal Election Institute (IFE), Dr Jorge Carpizo, claimed that the country should be proud of delivering the most open and systematized elections in its history, a view acknowledged by most national and international observers (Carpizo, 1994). In a speech that rejected claims that the election was fraudulent, Carpizo cited a compendium of statistics showing the scale and implied credibility of the task. According to the data, prior to the elections voter lists with the details of almost 46 million Mexicans were subject to 36 separate audits, were posted in town halls and information stands for cross checking, and verified by political parties and citizen groups. At most, the lists showed a 3.9% inconsistency and few 'ghost' voters. On election day, over 385,000 trained 'citizens' managed the polling stations, a quick counting method gave initial results for almost all the 96,394 voting booths, these results were checked 25 times, the voting was monitored by 81,620 national and 934 foreign observers, and over 6,000 stations were subject to random checks from the IFE. Carpizo praised the role of NGOs, especially Alianza Cívica, in the monitoring of the election process and, in an odd choice of metaphor, claimed that the IFE's procedures were 'building a new buffer [colchón] of truthfulness' (Carpizo, 1994: 5).

Research has, however, identified numerous constraints to the consolidation or deepening of the 'universal' democratization. Among the most commonly cited constraints are opposition from paramilitary groups and drug gangs, the instability caused by the geopolitical interests of the United States, the re-emergence of populism and the self-coup (*auto-golpe*), growing abstentionism, over-centralization of mandates and budgets, corruption and a politicized judiciary (Ambos, 1997; Cameron, 1998; Grugel, 1999; Philip, 1999; Tedesco, 2000). Reflecting the sense of an interrupted teleology, a vernacular vocabulary has emerged referring to the democracy on offer as imperfect, stalled, post-dictatorship, *democratura*, or *dictablanda*. Yet, few studies seem concerned to conduct ethnographic work to find out why so many people appear ambivalent to the virtues of democracy and willing to sacrifice ideological struggles for anomie, personal safety and 'shopping' (Colburn, 2002). The preference is to find solutions to constraints through self-reinforcing institutional scenarios in which civic participation, under the banner of good governance, can deepen the respect for democracy as well as achieve economic 'fitness' through democracy (Blair, 2000; Philip, 1999; Schönwälder, 1997; Tendler, 1996). One can witness an almost unseemly competition to identify 'innovations' that promise new ethical-political principles and relationships between public and private such as the civil society forums in South Africa, Panchayats in India and participatory budgeting in Brazil. While de Sousa Santos (1998) argues that participatory budgeting has marked a shift from techno-bureaucracy towards techno-democracy, he and others have noted that this shift has been difficult and uneven (Abers, 1998; Baiocchi, 2001). Moreover, pressures to emulate the eventual success have raised the danger of innovations becoming formulaic exercises that can be installed in an ideal sequence without much understanding of the original cultural specificity. As Avritzer (2002) notes in a stimulating critique of transition theory, the institutional perspective is unable to bind culture into the design of neat generic political institutions. I will return, briefly, to participatory budgeting later in the chapter.

Parallel to the concerns for the quality of deliberative democracy have been misgivings about the quality of citizenship. According to Pinheiro (1996), we can witness democracies without citizenship. In the words of O'Donnell (1999), there is the threat of 'low-intensity citizenship', or in the view of Holston and Caldeira (1998: 263) of 'uncivil democracies' caused by a 'disjunction' between the successful democratization of political institutions and the limited changes to the civil components of citizenship. Broadly, there is a concern that Latin America has broken the sequential evolution of rights set out by, for example, Marshall or Habermas (see Marshall, 1998; O'Donnell, 1999). In the Marshall map, citizenship would be bestowed through the state in a sequence from civil to political, and eventually to social rights, as marked out by the extension of property rights and the protection of individual freedoms by the rule of law, to involvement in government and its mandate, and eventually to the assignation of universal welfare. In Latin America, a series of liberal constitutions based on US and French templates had established a normative case for social equality through proscribing a series of civil rights and social entitlements, with the latter manifest in the urban landscape by the 1960s as public housing estates, clinics, schools, universities, pension fund and trade union buildings. The conferment of these social rights, however, was not dependent upon political rights in the shape of democracy, but closely associated with populism, clientelism and authoritarianism, and was culturally dependent upon recipients' conforming to the modern state project (Auyero, 2001; Gay, 1999; Roberts, 1996). Then, just as political rights and to some extent civil rights seemed more in evidence, so social rights were threatened by structural adjustment programmes within which are embedded a conceptualization of social rights based upon the supremacy of the private sphere over the public, that private rights realize private social needs, and that the intervention of the public (through the state) is harmful to the general good. Studies have since demonstrated the undermining of citizenship manifest as social polarization, increases in the percentage of people living in poverty, and 'new' forms of marginality as people from previously 'consolidated' communities become 'un'-employed, -educated, -serviced, and express social tensions through drugs, crime and gang culture (Auyero, 2001; Katzman, 1997; Ocampo, 1998; Pereira Leite, 2000; Robinson, 1999; Rubio, 1997; Zaluar, 2001).

Furthermore, despite the institutionalization of political rights, there is mixed evidence that civil rights are guaranteed by the epithet of a democratic legal state. The perception is of laws being cast in order to reflect the private interests that have 'colonized the state' rather than as an expression of a consensus of private rights and public obligations. Consequently, the law's normative value is diminished and everyone not served by its particularistic formation ignores the rule of law. As highlighted by human rights abuses, many of which are conducted by the police and legitimated by politicians who condone 'rough justice', the state itself transgresses the civil rights that it has set out (Caldeira, 2002; Davis and Alvarado, 1999; Holston and Caldeira, 1998; Pinheiro, 1996). As Caldeira (2002: 257) observes, 'citizens may have substantially increased their awareness of rights and citizenship, but still evaluate justice and civil rights in terms of privilege and find that police violence – even vigilantism and lynchings – are reasonable alternatives for dealing with increasing criminality'. There is a good case that this misrule of law may have been extended by democratic change, as opposition victories undermined the long-standing patronage of political bosses who had instilled 'discipline' by granting access to state resources, and by 'good governance' that has threatened the police with greater transparency and 'performance' but not tackled issues of poor pay and promotion (Colburn, 2002; Davis and Alvarado, 1999; Holston and Caldeira, 1998; Tedesco, 2000). The lesson from Latin America would appear to be that while the ideal formulation presents democracy and citizenship as related and positively constitutive, and indeed political rights can become increasingly embedded as the result of the exertions of citizens, civil citizenship is highly contingent and may even be undermined by a democratization that threatens the social institutions that previously restrained police and political violence. There is an irony that the shorthand term for violence with impunity, 'Colombianization', refers to one of Latin America's most established democracies.

Concerns for the qualities of democracy and citizenship have focused attention on the expectations placed on civil society to the broadening of the public sphere (Calhoun, 1993: 276). Idealized by Habermas (1996: 29) as 'a far-flung network of sensors' that 'react to the pressure of society-wide problematics and stimulate influential opinions' beyond the established positions of political society and self-interested individuals, researchers were able to provide empirical support that the public sphere was becoming a site for communicative action for what Avritzer (2002) calls 'participatory publics' that speak to a variety of communities in meaningful ways (Lara, 1998; Shefner, 2001). Social movements provided a critique of the universality of democracy, demonstrating how 'citizens' were being excluded from possessing the rights of full citizenship. Movements also recast the linear narrative of political society as class-based and stressed notions of difference constructed through a politics of identity around new subjectivities, especially relating to ethnicity, gender, sexuality and religion, and drawing upon a concern with the everyday practices of survival through issues such as health care and education, domestic violence, consumption, alcoholism and drug rehabilitation (Alvarez et al., 1998; Escobar, 1992; Guidry, 2003; Harvey, 1998; Radcliffe, 1993; Schild, 1998; Slater, 2002; Stephen, 2001). Overall, movements appeared to demonstrate how 'articulating good reasons', to borrow from Benhabib (1996), could open up the politicization of issues not circumscribed by 'received' discourse in structurally unequal societies where 'reasoned deliberation' is not free and equal.

In this process the opening of print and television media and the development of electronic communications have created opportunities to construct 'new' narratives of social problems and political solutions (Avritzer, 2002). Hopenhayn (2002) is optimistic that more open communication has re-territorialized citizenship around issues of gender, race, sexuality and consumption, and Blacklock and MacDonald (1998), Fonseca (1998), Routledge (1998), and Stephen (2001) show how engagement with globalized discourses of human rights has provoked discussions of morality and justice that have shifted the discursive terrain away from the state. Lara (1998) notes how the debates and exposé journalism of magazines such as *Proceso* supported the persuasive narratives of social movements in Mexico. By contrast to 1968 when massacres of students in Mexico City was given almost no media coverage, the 1994 Zapatistas uprising was 'imagineered' and keeping the EZLN 'in the news' was an important safeguard against repression (see also Routledge, 1998). Differently, others have stressed the importance of forums and commissions for truth and reconciliation as protected spaces for communicative action, in particular through the device of 'testimonio' (Fonseca, 1998; Nolin Hanlon and Shankar, 2000; Stephen, 2001). Exemplified by the commissions, above all, is an appreciation that the 'uncivil' discursive positions of the state or reactionary groups form a part of a deliberative democracy, but that such positions will appear unreasonable against the normative appeal of a society that people want to live in and therefore deny their claim to a universal validity.

Nevertheless, the contribution of social movements qua civil society has been subject to critique, posing some awkward questions about the hitherto axiomatic associations of democratization, civil society and social inclusion. Becker (1996) and Pearce (1997), for instance, observe that most writers have been content to normatively define civil society as a 'good thing' and implicitly represent it as free-floating from political society. One consequence is that writers have paid too much attention to the kind of civil society they like, ignoring those elements involved with a range of 'incivilities' or undemocratic behaviours, or expressing explicitly pro-capitalist aims (Becker, 1996; Brysk and Wise, 1997; Lucero, 2001; Pearce, 1997; Rubio, 1997). A second concern is methodological. Despite claims of being based on ethnography, the literature provides few first-hand accounts of how people construct identities and engage with movements. Consequently, perhaps, it has been possible to represent movements as part of civil society in ways consistent with a Rawlsian social contract in which groups formed for very specific interests are perceived to work cooperatively, with reasonable and moral intent, for a citizenship that is fair to all. A wide variety of movements have hence been depicted as in conflict with the state and not with each other in the pursuit of individual self-interest within organized structures (Becker, 1996; Pearce, 1997; Roberts, 1997). Generality has even permitted writers to ascribe certain movements as celebrating an identity politics of difference as embedded within a cultural context, without obliging them to say how this cultural politics of resistances needs to be (problematically) spatialized and why their pragmatic agendas are much less progressive. In the case of the Zapatistas, for example, while numerous representational devices were employed to re-appropriate national symbols as belonging to the poor and indigenous, and not the property of the state, the movement's proposed alternatives to neo-liberalism are quite close to the social citizenship enshrined in the liberal constitutions of the nineteenth century that were hostile to indigenous identities in the name of 'one nation' (Gledhill, 2002; Harvey, 1998). The imagineering made 'Chiapas' a public space for the representation of many peoples' struggles, yet perhaps it has also lost some strategic weight. The 'public' Chiapas has become an eco-Left chic heritage site in which the years of struggle can be captured in a three-page Sunday supplement or 20-minute documentary, that often times come close to repeating the discourses of those who sought to undermine the movement. The EZLN is represented as politically naïve, fighting a lost cause, because globalization and democracy in the progressive 'Third Way' has moved on.

A third, and related, criticism is how during the 1990s many movements began to operate with particular interests in mind, transforming themselves into non-governmental organizations, or what Schild calls neo-governmental organizations, sometimes at the direct service of the state. As such, movements' much praised autonomy has become somewhat less the republicanism of Rousseau and rather more a Friedman-esque 'freedom from' interference and assistance of capitalism (Calhoun, 1993). Furthermore, in re-representing themselves as viable providers of 'public' services, the politics of resistance may have been sacrificed for operation within the 'rules of the game' and narrower pragmatic agendas (Gideon, 1998; Gledhill, 2002; Pearce, 1997; Schild, 1998). In so doing, some have adopted the actions and discourses of 'routine politics' or been demobilized by political parties, the church, international donors and other NGOs. Movement leadership has become more professional, membership has remained high but mobilization has slackened off, and at meetings people speak longingly of a 'glorious past', although this social memory is an important resource for identity formation (Guidry, 2003; Roberts, 1997; Shefner, 2001). In practical terms, some studies suggest that the enthusiasm and innovative approaches of movements qua NGOs have not translated into improved social equality or in stemming the growth of poverty across Latin America (Robinson, 1999; Schönwälder, 1997; Shefner, 2001).

Social movements, as well as wider conceptualizations of civil society, have not proven to be the unproblematic vanguard of 'alternatives to development', ushering a radical transformation of ideas of modernity and society-economy (see Escobar, 1992). Rather, some regard the 'can-do' attitude of civil society as complicit with the limited interpretations of 'good governance', the consumer-citizen ideals of 'Third Way' populism, and the development discourse that denies even the promise of egalitarian rights-based social citizenship and elides civil society with the nurture of social capital (Gledhill, 2002; Portes, 1998). While the concept of social capital has put the 'social' back on to the development agenda, it idealizes a civic republicanism in which political duties are minimal (or unclear) unless functional to economic or social well-being (Molyneux, 2002). Moreover, social capital also presents notions of community and networks of trust that stress the virtue of mostly small-scale and local, and loosely bound forms of organization, but without acknowledging that many of the 'best practice' forms of social capital are constructed around organizations set up by the Left, liberation theology focus groups, women's unions and other movements (Molyneux, 2002; Portes, 1998). By building upon social organizations and institutional arrangements that are already in place, social capital seems to offer few opportunities to construct new forms of interaction, and without a theory of communication, the assumption that the outcome will be a radical democracy is not entertained (Putnam, 1995). This leads to the possibility that civil society's engagement with marginalized groups as a substitute for the state has re-marginalized these groups by, simultaneously, denying them the right of equal access to state institutions and obliging the poor to give up the leverage of clientelist vote-trading in return for an inspiring Third Way discourse of rights or the do-it-yourself of social capital formation (Gay, 1999; Shefner, 2001).

Finally, there is concern that the debates and networks in the 'global public sphere' often appear vague or disembedded from local conditions, and that the universal rights discourse consists of predominantly Western social and cultural values that offer few signposts to an understanding of difference and limited opportunities for the South to influence discursive content (Blacklock and MacDonald, 1998; Gledhill, 2002; Grugel, 1999).[1] Thus, in order to engage with global civil society and media, groups have to appear different but also to fit in with the universalizing norms and discourses, thereby obliging highly contextual interests to be represented as general and 'acceptable', and for fluid and multiple identities to be essentialized into stable and 'politically correct' categories understood by wider audiences. Hardly surprising that some movements have found it easier to transnationalize rather than work out how to relate universal discourses to local conditions (Robinson, 1999) or that some have misjudged how groups will engage with global discourses (Shefner, 2001). Others, however, have shown considerable skill at addressing these tensions by devising an external representation of essentialized unity (as women, indigenous, poor) and negotiating internally their identities and strategies, hybridizing the new vocabulary (Gledhill, 2002; Stephen, 2001).

Democratic Opening, Uncertain Citizenship, Fragmented Cities

The idea that citizenship may be impoverished just as democracy appears to have, institutionally, become more secure has raised questions about the grounding of citizenship in the nation-state. As Holston and Appadurai put it, there are new uncertainties 'about the community of allegiance, its form of organization, manner of election and repudiation, inclusiveness, ethical foundations, and signifying performances; uncertainty about the location of sovereign power; uncertainty about the priorities of the right and the good; uncertainty about the role of cultural identities increasingly viewed as defining natural memberships' (1996: 184). In place of the nation-state, Holston and Appadurai suggest that cities are becoming the space of citizenship, 'as the lived space not only of its uncertainties but also of its emergent forms' (1996: 185).

This is a more specific diagnosis than that offered by social scientists, one that ascribes a vibrant public sphere to the presence of accessible public space configured as

> the space 'where freedom can appear'. It is not a space in any topographical or institutional sense: a town hall or a city square where people do not 'act in concert' is not a public space in this sense. But a private dining room in which people gather to hear a Samizdat or in which dissidents meet with foreigners become[s] public spaces; just as a field or a forest can also become public space if they are the object and the location of 'an action in concert', of a demonstration to stop the construction of a highway or a military base, for example. These diverse topographical locations become public spaces in that they become the 'sites' of power, of common action co-ordinated through speech and persuasion. (Benhabib, cited in Allen, 1999: 211)

Similarly, Young's (1996) vision of an un-oppressive society is predicated upon a variety of public spaces that are accessible and where people can engage face to face, simultaneously aware of each other's presence, and yet co-exist as strangers, sometimes chatting, storytelling or debating with one another. Yet, if these qualities are to be associated most strongly with cities, to what extent do we witness in Latin America, as in Europe and the United States, a rupture between the dynamism of the public sphere and the accessibility of public space?

Are city public spaces threatened such that they no longer serve as 'open minded' arenas where diverse sets of people meet and engage in an uninhibited civic culture but have become 'single-minded' spaces, sterile, mediated and 'empty', subject to management and proscription of behaviour (Sennett, 1977; Walzer, 1986)?

There is evidence that the disjunctive citizenship described by Holston and Caldeira (1998) rests on a spatial fragmentation, in which peoples' search for a substantive content of citizenship relies upon a series of exclusionary turns. Across the region, middle and upper income groups represent themselves as 'new marginals' forced to isolate themselves behind the physical walls of gated communities, regulations and the imagined walls of 'taste' (Caldeira, 1996; Colburn, 2002; Coy and Pöhler, 2002; Kaztman, 1997). In Buenos Aires, Caracas and Santiago, business-resident coalitions have used decentralization initiatives to dis-incorporate elite areas from municipal government, impose restrictive ordinances on street traders and prevent property taxation being used to cross-subsidize investments in other, poorer, areas of the city. Civil association and NGOs with links to discreet residential areas or real estate companies have promoted environmental discourses to re-zone areas to lower densities, and therefore better-off social groups, and condominium boards set out obligations to join simultaneously golf, polo and country clubs in order to buy into certain communities. Services to elite communities, for maids, nannies and gardeners, are increasingly mediated by contractors, removing even the paternalistic contact with 'servants', and many of the larger developments include malls, schools and hospitals within the walls, and new 'civic' amenities such as art museums and private universities located adjacent. In 'selling' places developers are picking up on peoples' concerns for security by stressing the spatial isolation and separation of new developments; casting spatial distance as a sign of social distinction (Caldeira, 1996). One consequence is that spaces for heterogeneous and spontaneous contacts diminish (Caldeira, 1996; Katzman, 1997), especially as peoples' response to fear of crime has been to identify definite routes between places and minimize the probability of *ad hoc* detours.

To García Canclini (2001) these trends mark a shift in Latin America's urban aspirations towards the fulfilment by 'non cities' where exclusion is reinforced by the creation of 'public spaces' in which people are physically proximate but disengaged other than through a private ethos of consumption. Citizenship, he argues, is experienced in spaces that have lost their normative focus such that it is no longer possible to imagine the construction of a modern urbanism of Paris, New York or even Miami but only 'disintegrated' cities that have become the 'suburbs of Hollywood'. In Puebla, Mexico, the new malls that replace the 1970s and 1980s versions located within the city, anchored by a supermarket and cinema, and potentially accessible to a broad range of people, are accessible only by car, with the parking serviced by rapid response security teams and CCTV, and costing many times a daily minimum salary. The malls themselves are designed as closed-off spaces with defensive architecture, only a half dozen entrances, climate controlled, no toilets, and few public phones or seats. Their 'public spaces' include attractions such as ice rinks and display spaces for imported cars, especially sports models and 4×4, while typical stores will be The Noble House or Cartier. During visits, it is rare to see indigenous or darker mestizos, even as security guards or shop workers: in the Angelópolis mall the only exception were servers at a MacDonald's restaurant tucked into the corner of the food hall. The malls offer a natural catwalk for the *rimbombante* youth to be actors and spectators, but in the Angelópolis mall I could not find a bookshop. Policed and disciplined by the architecture, location and clientele, the gated and mall communities are more predictable than the city outside or the US or European cities that the developers profess to mimic.

Spaces beyond these gated 'communities' also appear to be symbolically more private as representing the consumption values of particular classes. In Puebla, again, the 'recapture' of the downtown through conservation and conversion into a 'centro histórico' stresses an elite, Europeanized and whiter representational space and greater order, cleanliness, morality and dignified use (Jones and Varley, 1999). In contrast, perhaps, to the loss of confidence in public space described by Sennett (1977), related to the increasingly plural city that made dress/appearance/demeanour/language less reliable signals to status, Latin American cities have been historically far less cosmopolitan (Rosenthal, 2000). Elites and middle classes are acutely cognizant of social distinction, but their concern is less that differences are blurred in public space but that such differences remain all too obvious and anxiety-inducing. Some argue therefore for the removal of the poor, loud, dirty, indigenous, traders, youth, who exhibit their difference from the normativity of tourist brochures or lifestyle magazines by how they eat, play music, swear or their inability to consume. This public must be removed from certain public spaces because they 'are not what they once were' and need to be recaptured for 'the' public. The new public spaces in the centres of Puebla, Quito, Cartagena, Mexico City, Recife and Rio de Janeiro are art museums and convention centres, European-style coffee shops, or recreations of past landscapes such as the proposed Opera House on the *alameda* of Puebla, a space regularly occupied by fairs, street traders and demonstrations (Jones forthcoming; Streicker, 1997). In Bogotá the Cultura Ciudadania (Cultural Citizenship) campaign of mayor Antanas Mockus has removed street traders in order to create designated spaces for civic participation while in Buenos Aires, media-savvy politicians have appropriated public spaces to enliven dry technocratic campaigns and attempt to convey the public nature of a politics that stresses an increasingly private ethos (Waisbord, 1995). What or who is allowed to remain in the centre is partially defined according to a position in a particular, private, historical narrative.

With the possible exception of Brazil, the democratization of Latin America has not been accompanied by a debate about what type of vision of urban space is compatible with more inclusive notions of citizenship. How does the mall or gated enclave engender citizenship as a private public space with few opportunities for demonstrations or graffiti? Elites, especially, wired into global political and consumption networks seem to be living in an ever more reduced physical and metaphorical 'politics free zones' (see Colborn, 2002). Should we be concerned that *voluntary* membership of Blockbuster video now probably exceeds that of all major political parties? How can norms be constructed where difference is defensive, diversity is mistrusted, and education, even if it had rarely ever meant learning *with* 'others', now might not impress to learn about others either? What kind of radical or deeper democracy can emerge in societies in which a numerically small but economically significant group is physically isolated, socially less engaged and economically less dependent on the remainder of society?

The Geopolitics of Democracy and Citizenship

These are important questions. But the image of elite groups privatizing space against poor 'others' risks overdetermination of the spatiality of power. In gating themselves against the social realities outside or in constructing powerful discourses to exclude 'others' from public spaces, elites represent themselves as vulnerable and marginal, but are represented in the literature as effectively powerful. Yet, social power does not begin or end with spatial distance or socio-economic polarization. If we imagine power as diffuse and communicated rather than held and transmitted we might see cities as rather different sites of citizenship, possibly disjunctive nonetheless (Allen, 1999). Consider, the following vignettes.

The first considers that a young, possibly black or indigenous, poor, and most probably male, person's presence in a certain part of a city transgresses numerous lines of tolerance and discipline, and might mobilize security concerns. Yet, the power to define being in 'the wrong place at the wrong time' is not the sole preserve of better-off groups or the state. In early 2003, the Minister for Urban Development in Brazil, Olívio Dutra, visited the *favela* 'New Brazil' in Niteroi near Rio de Janeiro. Marking the Lula administration's intent to include the excluded, the symbolism of the event also revealed that Dutra had to request permission from the leader of the local gang, Comando Vermelho, who at the time was in prison. A youth nearing a gated community possesses a power to illicit fear and a response, the Minister for Urban Development and all his bodyguards elicits little fear from the *favela* (on this occasion) and requires permission to enter. In power relations the *favela* should be seen as 'gated'.

Second, a few years ago I was standing on the running board of my Volkswagen Beetle taking some photos of the Santa Fe shopping mall and commercial centre when I was arrested by two policemen. My interest in Santa Fe was as a site, partially built by Olympia and York on a former rubbish dump, that from its opening in 1994 had become the symbolic epicentre of a globalized, NAFTA-inspired postmodern Mexico. The police claimed to be guarding Santa Fe and were suspicious as I could have been 'casing the joint' for a robbery. My argument was that even if I were the mastermind of a crime syndicate they could not detain me as I was, technically, either on private property (my car) or federal property (the highway), and was therefore beyond their jurisdiction. Without any idea whether this argument was correct, I was sure that I had committed no crime, but geography seemed a more awkward defence than legal details. The police wanted to impound the car, but then decided to take my camera and, finally, just confiscate the film. At this point my companion intervened to explain that maybe this could all be cleared up by calling her boss. For ten minutes neither policeman had shown any interest in Susana but once she explained that her boss was the Head of the Chief of Staff at Los Pinos (presidential residency) and then produced her ID to indicate the rank of army captain, attitudes changed. What in Mexico is called a *charolazo*, the use of symbolic influence to overturn power relations, duly produced my camera and a complimentary police escort for our journey. The *charolazo* and not the rights of citizenship transformed my being out of place to becoming in place, and legitimated my access to Santa Fe as public space over the owners' attempts to use the police as private security and the highway as private space.

Both these vignettes suggest to me a need to avoid ascribing particular political attributes to spatial referents, and how we might think about multiple spatialities of politics by reconsidering the notion of the 'public' space. Generally we have rather hackneyed characterizations of the boundaries between public and private, which appear to be no different in Europe, from the USA, from Guatemala to Uruguay, from Coatzacoalcos to Buenos Aires. Not only does this impoverish the writing on the Latin American city, but also it tends to deal uncritically with cultural distinctions between public and private, as well as vest power in identifiable agents rather than those who resist the distinctions between private and public. Clientelism, for example, remains embedded in Latin American politics, extending personal social relations into the public realm and serving as a relational form of citizenship more meaningful to many individuals than the abstract versions set out by laws (Auyero, 2001; Gay, 1999; Guidry, 2003). Fortunately, some researchers are theoretically and empirically beginning to expose how peoples' social practices have profound effects upon the contest for the multiple representations and meanings of space that involve challenging the boundaries between the private and the public (García Canclini, 2001; Lewis and Pile, 1996; Rosenthal, 2000; Streicker, 1997). My argument is, recalling Benhabib's comments above, that we should consider public spaces as the sites of the contestation of power. Spaces are made public when they inform others about the discourses of the groups that physically occupy them or symbolically invest them with meaning. Public space therefore is performative, where identities are exposed and communicated, interpreted, understood and transformed. Spatial representations become strategic sites for contests of democracy and citizenship.

In proposing this argument I am drawing upon the work of Avritzer (2002), Melucci (1996), and Mouffe (1993) to re-examine the relationship between social movements, politics and the public sphere. In particular, I want to put less emphasis upon movements or other groups as institutions, and more upon the discursive engagement between politics as the practices, discourses and institutions that order social life, and the political which is the disruption of these orders (Slater, 2002). One such disruption occurs when expressions of identity are presented in public as different from the normative order. To Mouffe, the universal or public sphere is viewed as a discursive space and it is the decisions on how to fill this space that defines the act of politics. As Harvey explains: 'the reality of citizenship is transformed from a pre-given "signified" (which has only to be realized through political reforms), into an "empty signifier" whose content is filled by the political choices of social agents' (1998: 160). Citizenship, therefore, is non-determined, allowing for a constant process of contestation as identities are worked out in relation to an imagined 'public' (Young, 2001). The disruption reflects points of resistance to the imposition of order, therefore setting the political within a network of power relations that can never be resolved towards a single position (Slater, 2002). Here, then, citizenship has no defined meaning and cannot be thought of as a concession from the state, but is actively and constantly constructed through identification to a variety of groups or system of values that are not based upon specific pre-given bounded identities.

I want to develop these ideas by suggesting that we need a broader conceptualization of communication than that usually provided by work on the public sphere, which retains a strongly formal and institutional perspective. As Young (1996) has argued, deliberation should include greeting, rhetoric and storytelling, and that a vibrant democracy is likely to be rowdy and disorderly. Unfortunately, our attention to civil society and social movements in Latin America has provided few ethnographies of how identities are worked out and presented publicly through more vernacular forms of communication. Yet, where such ethnographies do exist, they reveal that the institutional failure of a movement, while lamentable in its own terms, was not the whole story (Jones, 1998). My argument is that the process of making identities public employs the use of a range of forms of communication, beyond those even of Young, to include cartoons, humour and anecdote. These often provide contentious representations of spaces as 'public' in particular ways in order to contest the legitimacy of state or other actors. In the remainder of this chapter, therefore, I want to open up some of the implications for our understanding of democracy and citizenship suggested by multiple spatialities of politics.

The first example takes the idea that public spaces were 'lost' during authoritarianism as the state attempted to define the representations of space in order to limit the public itself. Public spaces became 'empty' spaces, squares were 'opened' with the removal of trees, benches, bandstands, to become huge concrete plinths occupied by the national flag and official ceremonies, while other spaces were built, such as football stadia and auditoriums where the public were spectators and the spaces symbolized public space but were not occupied freely by the public. Instead, in order to assert the primacy of the nation, spaces such as prisons, detention centres and army barracks, but also parks, squares, stadia, roadsides and garbage dumps, put on public display how those who failed to conform might 'disappear' to be mourned, officially, only in private grief

(Nolin Hanlon and Shankar, 2000; Radcliffe, 1993; Scarpaci and Frazer, 1993). Resistance to authoritarianism, therefore, attempted to retake spaces for the public through representation in films and murals, and physical occupation through marches, hunger strikes and sit-ins (Radcliffe, 1993; Scarpaci and Frazer, 1993; Waisbord, 1995). In Puebla, so frequent were the occupations of the main square during the 1980s and 1990s that it became known as 'La Borracha', literally 'the drunk', because it was always 'tomada', from the colloquial use of the verb 'to take' or 'to drink' (and 'to fuck'). In Mexico, authoritarianism is represented by Tlatelolco Square where in 1968 the student movement was ruthlessly put down. The symbolism of 1968 runs throughout the democratization of Mexico, and is often used as the 'date' from which the emergence of social movements and the opening of the media is deemed to have begun (Lara, 1998). Spatially, however, Tlatelolco's symbolic relationship to authoritarianism runs from before 1968, as the site of Aztec blood rituals, overlaid by the Spanish with the church of Saint James, and after, when the state denied local involvement in the redesign of the Le Corbusier-style 1960s 'superblocks' that were flattened by the 1985 earthquake. Tlatelolco has been 'made public' as the site of a massacre, as the site of syncretic nationalism, as the site of multiple social movements and as a place that appears in numerous political discourses in order to draw upon complex meanings of sanctity and sacrifice, modernity and dis-modernity, authoritarianism and democracy.

The second example is participatory budgeting in Brazil which, as Abers (1998), Avritzer (2002), Baiocchi (2001) and de Sousa Santos (1998) show, has created an innovative form of deliberative democracy. Most writing on the budget decision-making process, however, has concentrated upon its political innovation and the evolution of technical procedures to maintain enthusiasm, equality of voice, a fair distribution of resources, and prevent co-optation by civil movements or old-style politicians. What interests me is how participatory budgeting has re-spatialized politics, becoming a representational device to dramatize democracy. I think that one can understand this in a number of ways. First, budgeting changes the relationship between the people, politicians and bureaucracy, not only in weakening the ties of clientelism but, as Abers describes for Porto Alegre: 'on numerous occasions, budget participants flooded the city assembly chambers to demand that the deputies approve the budget they had designed and vote for tax increases that financed investments' (1998: 518).

Storming city hall is not new, but doing so to demand that a budget is set, one designed by the people after dialogue and which may involve tax increases, is a dramatic shift. Second, in conversations with municipal officials in Belo Horizonte and Recife I was struck by how indeterminate and different the language was to explain their attendance at neighbourhood meetings. They knew that resources would not match expectations and that certain works were more feasible than others, but how were they to talk with people whose concerns were not technical but emotional, who perceived services as a right and city hall's role as a responsibility. The respondents, by no means aloof or dry technocrats, saw that they had to engage in a dialogue rather than rely upon co-optation of leaders, the imposition of a party line or just closing the door and waiting for the problem to go away. City hall had to go out to the people, and it had to learn how.[2] Third, participatory budgeting has animated spaces within neighbourhoods. While we still know very little about how people engage with the budgetary process, and how the act of participating in public has changed beliefs in justice, citizenship or themselves, Abers (1998: 526–529) describes how at neighbourhood meetings numerous 'scenes' occur as people work out how to speak with one another in public, how to be silent, how to persuade or be persuaded. The drama of these discussions resonates long after the meetings. As I witnessed in one *favela* in Belo Horizonte, people who had attended a recent meeting were asked by neighbours about what happened and a space had been set up on which a summary of proceedings was posted and opinions solicited.

The third example reveals democracy as a performative act rather than just an institution that can be right or wrong. The example takes us back to Jorge Carpizo's assessment of the democratization of Mexico. As a series of technical signposts, Carpizo, and most commentators, missed the drama of the 1994 elections. Yet, when I stood outside a polling station in the Morelia, Michoacán, from just after daybreak until long after it was dark and the ballot boxes had been taken away, the polling station was a contested space for various performances on democracy. There was the large crowd that filled the street outside and a queue that stretched for over three blocks waiting to cast their vote. There was a buzz that the PRI, the party in power for almost 70 years, might actually be defeated and some concern about what might happen if it did lose; would it go gracefully, would the army become involved, how would the USA react? There were conversations about events in Chiapas, the fate of the Left candidate Cuauhtémoc Cardenas if he were to lose again and whether the Right would form a coalition with the PRI. Intense political debates were broken by talk of football, food from street traders, and swapping personal stories and occasional signs of boredom.

Another set of performances revealed a tension between the legality and legitimacy of democracy. The polling station was a 'special', formed in order that people could vote away from their place of residence. The special stations, however, possessed only 300 ballot papers but an hour before opening there were already 600 people queuing outside. By 11am all 300 papers had been used and at 12.45 the station was closed. The reaction of the crowd was a mixture of mock resignation and genuine anger. At this point two observers from the Santa Rosa catholic mission in California decided to intervene. With what appeared to be no knowledge of Mexican politics and only a rudimentary grasp of Spanish they suggested everyone ask guidance from God through a mass prayer. The observers were mostly ignored or became a source of humour for the crowd. While this performance went on, a section of the crowd formed a commission to request the intervention of the IFE and a short time later two officials arrived in a dark pick-up, dressed with reflector sunglasses and the self-presence of judicial police. The performance of these 'men in black' was to represent the rule of law, speaking calmly to the crowd and requesting more ballot papers from the IFE office. They tell the crowd that everyone has a right to vote who remains in line at 6pm. At 4pm, however, a representative on the IFE council (*consejo local*) arrives, and with a copy of the government gazetteer explains that 'special' stations are only permitted 300 ballots and that everyone who has not voted (approximately 600 people remain) should go to the only other special station in the city. The *consejero* was aware that public opinion was against such a narrow interpretation of law so to those within earshot he admitted that the second special station had finished voting long ago. An angry section of the crowd is told that they can send their complaints to the IFE in Mexico City. When, later, it is announced that the election attained a 78% turnout, the 600 people unable to vote in Morelia appear as abstentions. The only substantive admission of failure acknowledged by Carpizo was the special polling booths which, he notes, were limited to 687 stations nationwide with 300 ballot papers each by agreement with all political parties. To Carpizo, the polling stations were spaces for democracy, constructed by the state and protected by law. What Morelia demonstrates is that polling stations can also be understood as spaces of democracy in which political dramas and debates are performed. As such, they became a public space beyond the instrumental limits of democracy in which other spaces (in this case, the nation, Chiapas, home) are discussed as part of the public sphere. Democracy, therefore, has served to exhibit subjectivity and social interactions through the construction of a particular public space. Within this setting, a range of relations are built between individual and group subject positions and different spatialities. In order to understand the meaning of democracy, therefore, we need to be more attuned to the spaces in which politics takes place and to the spatialities of a political performance.

Citizenship and Democracy in a 'Failed' Movement

Finally, let me consider how, in regarding spatial representations as strategic sites for the contest of citizenship, we can challenge some of the notions that social movements have 'failed' in recent times. As I have done elsewhere, I want to consider a network of social movements that was explicitly formed in order to challenge the privatization of public space in both the physical and more discursive senses but which, from an objective or resource mobilization standpoint, failed (see Jones, 1998). I want to think about how resistance drew upon a series of spatialized discourses that animated a politics of identity and recast a sense of citizenship. The example starts with an announcement on 6 July 1993 by (then) Mexican president Carlos Salinas of the launch of a US$ 1 billion 'megaproyecto' to 'recover the grandeur' of Puebla through the 'detonation' of foreign direct investment. One sub-project was the creation of a 27-block zone covering four *barrios*, part of the historic centre but largely ignored by the conservation projects of the previous decade (Jones and Varley, 1999). The San Francisco project would deliberately alter the orientation, socio-economic base, architectural form and possibly ethnic composition of the *barrios*. At the heart of the project written by McKinsey and Company and later refined by HKS-Sasaki of Dallas would be the establishment of a cultural, tourist and business district that would integrate San Francisco to the 'Spanish' centre. The main road, Boulevard 5 de Mayo, which separated the *barrios* from the historic centre, would be modified to reveal the San Francisco river, canalized in 1963, and this would be made navigable to small launches to take people between two artificial lakes. In one version, the Riverwalk would link to a cable car system to take people up to the forts, and the lower site would include new museums, hotels, a library and convention centre.

The San Francisco *megaproyecto* represented the *barrios* as incompatible with the reconstruction of Puebla as a modern city. Moreover, as the historic centre had been renovated over the past 20 years, and as certain uses and users of that space had been removed and elite consumption practices reasserted, so the *barrios* seemed all the more 'out of place'. The perceived degradations of the *barrios* legitimated intervention to rescue the built heritage (*patrimonio edificado*) even though neither government nor the initial consultants conducted a 'ground truth' survey to consider whether their perceptions were in any way accurate – when a survey was conducted in 1997 less than one-third of buildings were classified as in poor or ruinous conditions. In documents and public discourse few mentions were made to the population of the *barrios*, other than graphically as old, post-industrial textile workers, possibly responsible for the conditions of the area, and in contrast to the aerial photography and neat architectural drawings representing the state's intentions. Conventionally regarded as 'Indian', the *barrios* now appeared not to be indigenous enough, commentators noting that few people spoke any indigenous languages and there was no distinctive dress. In a series of videos, lectures and, later, a book called *The Foundation and Development of the City of Puebla de los Angeles* during 1998, the themes covered were archaeological discoveries and the sixteenth- and seventeenth-century history of the city. The twentieth century was ignored except for a photo display of destruction and decline. Within these themes it was repeated that Puebla was founded through religious significance and the San Francisco project was described as a means to 'liberate' buildings and recover 'customs and traditions'.

The San Francisco project provoked a series of social movements. These included the Unión de Barrios that represented tenants, small labour organizations, street traders and artisan groups, most of whom the Unión was keen to prove lived in conditions of poverty, and the well-named Asociación Civil por los Ideales de la Puebla Tradicional that represented small property owners. These two groups united to form the Unión de Ciudadanos Libres (UCL), which allied to a host of other groups resisting other projects in the Angelópolis package, including peasants, residents in low-income settlements, parent organizations (a number of schools were being closed or destroyed as result of projects) and cultural lobby groups. Fighting the government discursively, the UCL failed to prevent the implementation of San Francisco. Ultimately, the state's original intention to amass 27 city blocks was reduced to a three-phase programme starting with the acquisition of six blocks, largely it was said because of budgetary shortfalls, investor confidence after the peso-crash and an opposition victory at the mayoral election. Nevertheless, the San Francisco-lite includes some small and mostly empty parks, an historical museum (now closed), a large convention centre, and some privately run restaurants and executive colonial-style condominiums.

What also matters, however, is how the movements challenged the discourses of a (imagined) cultural identity which was written for them and which represented the *barrios* as 'traditional' (but not indigenous), 'local' and 'popular' in the sense of being anti-modern. In a complex series of representations, the origins and deeper meanings of which I can only touch upon here, the UCL projected their identities through the multiple spatialities of the barrios, and in so doing sparked a series of debates that were taken up in the press and elsewhere. For brevity I will outline just three of the spatial representations around which citizenship was contested. The first was the idea of *barrio* identities as fixed in space. To the state the *barrios* were represented as having a fixed and knowable identity that could be dissected in order to retain certain elements for the spectacle of the San Francisco project and for others to be ignored. The UCL described San Francisco as 'cultural genocide' and likened the *barrios* to victims dying of AIDS (a distinctly postmodern image). They also represented the *barrios* as intangible social and cultural spaces in which identities were related to particular popular practices that were '*barrio*' but not reduced to single sites. So, for example, the UCL described and promoted the saint's days and the 'Procession of the Cross' (dating 1615), and practices such as the preparation of foods such as camotes, chalupas, and breads. These representations of *barrio* identities speak out to ideas of Mexicanness, and especially a Puebla identity (*poblano*). Yet, in order that these representations did not appear to mark the *barrio* (residents) as backward, superstitious or Indian, some food and religious-inspired events included classical concerts and highbrow talks about the history of the *barrios*. To Churchill (1999), the intangibility of this cultural identity was precisely the point; people associated with an ambivalent cultural context and considered it to be their 'right' to work it out.

The second representation was of the *barrios* as a site of collective opposition. Speakers for the *barrios* made frequent reference to Analco, one of the *barrios* and which means 'the other side of the river' in *nahautl*, or to the *barrios* as the imprisoned side, the side for the Indians that built the city, the oldest side, the most often flooded, the poorest, the most combative when it is necessary. In this *barrio* as opposition, the representation was of a space that has historically held out for national identity, and often proven right or victorious. Here, then, were the *barrios* now doing so against the insensitivities of NAFTA and globalization. As was often repeated in discussions, jokes and displayed in cartoons, the *barrios* were not 'Made in the USA', a point underscored by reference to the term in English, whereas Angelópolis was the product of 'foreign' consultants which made the programme 'undemocratic and unMexican'. UCL derided the Riverwalk as a pastiche of San Antonio, threatening to turn Puebla into a Disneylandia, Moll, Las Vegas or Texas city. More subtly, the involvement of French companies seeking to build new hotels in the *barrios* allowed UCL to represent themselves as descendents of General Ignacio Zaragoza, who defeated the French in 1863 with men from the *barrio*, an event commemorated in the Boulevard 5 de Mayo. Rather than modernizers, the government was scolded as *vendepatria* or prostitutes willing to sell out Mexico just as Santa Ana had sold Mexico to Texas, including San Antonio! Rather than accepting the representation of the *barrio* as unmodern, the UCL represented the San Francisco project as dismodern, as an excess of capitalist modernity that lacked originality, style and place. In various ways, through representations of the *barrios*, the UCL projected identities and visions for the *barrio* that were compatible with a progressive modernity – even citing Le Corbusier for inspiration!

The third representation projected the *barrios* as democratic spaces where rights were respected because civil associations were open and non-hierarchical. UCL organizations stressed the participatory nature of their organizations with officers elected every four months. The 'public' space of Chiapas emerges frequently in UCL representation and discourse, notably in using the symbol of a dove in a balaclava as the 'spokesperson' for the movement, and use of the familiar address of 'tu' (you) rather than the officious 'usted' adopted by government. An alleged interview and wishes of support with sub-commandante Marcos was disseminated in a newspaper distributed by the UCL. The *barrios*, moreover, acquired a spatiality that made them both synonymous with Chiapas and elsewhere and against the autocracy of government. The governor Manuel Bartlett was compared to a predecessor, Nava Castillo, who among a range of autocratic infrastructure projects (including the canalization of the Boulevard 5 de Mayo that also destroyed many *barrio* buildings) was also responsible for the deaths of students in a 'civil war' against the 13 de Octubre movement in the 1960s. Bartlett was linked to the assassination of Luis Donaldo Colosio (the presidential candidate allegedly killed by the Tijuana drug cartel) and known for his hard line on the EZLN. When, in May 1997 a number of houses were bulldozed at night, having been surrounded by between 500 and 800 police and while some of the inhabitants were still inside, *barrio* residents and UCL likened the democratic Mexican state to the Soviet Bloc where, it was especially noted, homes and churches were violated by the state (Churchill, 1999).

The discourses, verbal and graphical, from the UCL and others stressed how the *barrios* were different from the representations of the San Francisco project. This difference, however, was carefully constructed not to fall too comfortably into the neat oppositions of Indian–white, popular–elite, historic–modern, local–global which was partly the legitimation of the state project and which would, in most people's terms, represent both the *barrios* and its inhabitants as inferior. The UCL therefore attempted to blur these distinctions, making it problematic for the state to apply simple labels to opponents, and therefore making it difficult for the state to talk about the *barrios* with any sense of definition. UCL recast identities, grounding these and making them public through press releases, radio interviews, graffiti, and especially a free newspaper and cartoon series (Jones forthcoming). Through representations of the *barrios*, identities and citizenship were spatialized in multiple ways that gave meaning to the *barrios*, integrated into a city, nation and global economy seen by many as fragmented, resisting the 'democratic' state's intention towards the privatization of a public space through the 'publicity' of the *barrio*.

Conclusion: La Pachanga Politica

This chapter has looked for evidence of democratization and citizenship while attuned to the spatiality of politics. In this search, it is no great surprise that the literature on the democratic transition from an institutional perspective is mute on the importance of space. It simply does not matter so long as the 'institutions are right', bar perhaps a few comments about the uneven nature of democratic freedoms and the working of good governance. It is more revealing that despite reference to the distinction between the political and politics, that much of the literature analysing social movements and the politics of identity is not concerned with spatiality. This chapter has attempted to show that if we understand social practice to be discursive and the meaning of social action to be constantly changing and contested, then there must be a role for space (Massey, 1999). Specifically, I have looked at the city, noting that to many observers this space is increasingly fragmented. However, rather than read off from our perceptions of fragmentation particular power relations, my view is that power relations are far more messy. How people relate to and attribute meanings in spatial relations becomes a fundamental exercise of power. Therefore, contests for the representations of space serves as a locus for making political identities public. In making these spaces public, we get away from neat processes of privatization, as the 'publicity' of the spaces conveys views on society and polity – perhaps especially where formal deliberative democracy is circumscribed. The qualities and the limits to formal citizenship, therefore, become less neat and, even if the outcome is neither normatively nor conclusively optimistic, in so much as contestation contains the *possibility* that politics remains a project in Latin America, the 'pachanga política' (the game of politics) remains as dynamic as ever.

Notes

1   Attention to language is revealing as key terms such as empowerment, governance, participation or grassroots either have no ready translation into Spanish, Portuguese, or indigenous languages, or have multiple and ambivalent meanings compared to English.

2   We know very little about how participation has extended the public sphere in ways not technically framed by budget participation. Designers of the budgetary process have responded to demands for inclusion into decision-making of cultural and leisure activities, and environmental matters, that were not original considerations (de Sousa Santos, 1998). But participatory budgeting has no formal leverage on police or land tenure legalization which fall under the remit of state governments. I am grateful to Celina Souza at University of Bahia for confirming this point. The Lula administration is rapidly decentralizing tenure legalization.

References

Abers, R. (1998) 'From Clientelism to Cooperation: Local Government, Participatory Policy, and Civic Organizing in Porto Alegre, Brazil', *Politics and Society*, 26: 511–537.

Allen, J. (1999) 'Spatial Assemblages of Power', in D. Massey, J. Allen and P. Sarre (eds), *Human Geography Today*. Cambridge: Polity Press, pp. 194–218.

Alvarez, S., Dagnino, E. and Escobar, A. (1998) 'Introduction: The Cultural and the Political in Latin American Social Movements', in S. Alvarez et al. (eds), *Cultures of Politics, Politics of Cultures: Re-visioning Latin American Social Movements*. Boulder, Co: Westview Press, pp. 1–29.

Ambos, K. (1997) 'Attempts at Drug Control in Colombia, Peru and Bolivia', *Crime, Law and Social Change*, 26 (2): 125–160.

Auyero, J. (2001) *Poor People's Politics: Peronist Survival Networks and the Legacy of Evita*. Durham, NC: Duke University Press.

Avritzer, L. (2002) *Democracy and Public Space in Latin America*. Princeton, NJ: Princeton University Press.

Baiocchi, G. (2001) 'Participation, Activism, and Politics: The Porto Alegre Experiment and Deliberative Democratic Theory', *Politics and Society*, 29 (1): 43–72.

Becker, D. G. (1996) 'Citizenship, Equality and Urban Property Rights in Latin America: The Peruvian Case', *Studies in Comparative International Development*, 31 (1): 65–95.

Benhabib, S. (1996) 'Toward a Deliberative Model of Democratic Legitimacy', in S. Benhabib (ed.), *Democracy and Difference*. Princeton, NJ: Princeton University Press, pp. 67–94.

Blacklock, C. and MacDonald, L. (1998) 'Human Rights and Citizenship in Guatemala and Mexico: From "Strategic" to "New Universalism"?', *Social Politics*, 5 (2): 132–157.

Blair, H. (2000) 'Participation and Accountability at the Periphery: Democratic Local Governance in Six Countries', *World Development*, 28 (1): 21–39.

Brysk, A. and Wise, C. (1997) 'Liberalisation and Ethnic Conflict', *Studies in Comparative International Development*, 32: 76–104.

Caldeira, T. (1996) 'Fortified Enclaves: The New Urban Segregation', *Public Culture*, 8: 303–328.

Caldeira, T. (2002) 'The Paradox of Police Violence in Democratic Brazil', *Ethnography*, 3 (3): 235–263.

Calhoun, C. (1993) 'Civil Society and the Public Sphere', *Public Culture*, 5: 267–280.

Cameron, M. (1998) 'Latin American Autogolpes: Dangerous Undertows in the Third Wave of Democratisation', *Third World Quarterly*, 19: 219–239.

Carpizo, J. (1994) Some Comments on the 1994 Federal Electoral Process in Mexico: A message given on 12 September 1994. Mexico City, Secretaria de Gobernación.

Churchill, N. E. (1999) 'El Paseo del Río San Francisco: Urban Development and Social Justice in Puebla', *Social Justice*, 26 (3): 156–173.

Colburn, F. D. (2002) *Latin America at the End of Politics*. Princeton, NJ: Princeton University Press.

Coy, M. and Pöhler, M. (2002) 'Gated Communities in Latin American Megacities: Case Studies from Brazil and Argentina', *Environment and Planning B: Planning and Design*, 29: 355–370.

Davis, D. E. and Alvarado, A. (1999) 'Descent into Chaos? Liberalization, Public Insecurity, and Deteriorating Rule of Law in Mexico City', *Working Papers in Local Governance and Democracy*, 1: 95–107.

Escobar, A. (1992) 'Imagining a Post-Development Era? Critical Thought, Development and Social Movements', *Social Text*, 31: 20–56.

Fonseca, M. (1998) 'The Transformation of the Guatemalan Public Sphere: 1950s to the 1990s', *Social Politics*, 5 (2): 188–213.

García Canclini, N. (2001) *Consumers and Citizens: Globalization and Multicultural Conflicts*, Minneapolis: University of Minnesota Press.

Gay, R. (1999) 'The Broker and the Thief: A Parable (Reflections on Popular Politics in Brazil)', *Luso-Brazilian Review*, 36 (1): 49–70.

Gideon, J. (1998) 'The Politics of Social Service Provision through NGOs: A Study of Latin America', *Bulletin of Latin American Research*, 17: 303–321.

Gledhill, J. (2002) 'Some Conceptual and Substantive Limitations of Contemporary Western (Global) Discourses of Rights and Social Justice', in C. Abel and C. Lewis (eds), *Exclusion and Engagement: Social Policy in Latin America*. London: Institute of Latin American Studies, pp. 131–147.

Grindle, M. S. (2000) *Audacious Reforms: Institutional Invention and Democracy in Latin America*. Baltimore, MD: Johns Hopkins University Press.

Grugel, J. (1999) 'Development and Democratic Political Change in the South', *Journal of International Relations and Development*, 2: 403–414.

Guidry, J. A. (2003) 'The Struggle to be Seen: Social Movements and the Public Sphere in Brazil', *International Journal of Politics, Culture and Society*, 16: 493–524.

Habermas, J. (1996) 'Three Normative Models of Democracy', in S. Benhabib (ed.), *Democracy and Difference*. Princeton, NJ: Princeton University Press, pp. 21–30.

Harvey, N. (1998) 'The Zapatistas, Radical Democratic Citizenship, and Women's Struggles', *Social Politics*, 5 (2): 158–187.

Holston, J. and Appadurai, A. (1996) 'Cities and Citizenship', *Public Culture*, 8: 187–204.

Holston, J. and Caldeira, T. P. R. (1998) 'Democracy, Law and Violence: Disjunctions of Brazilian Citizenship', in F. Aguero and Stark, J. (eds), *Fault Lines of Democracy in Post-Transition Latin America*. University of Miami: North-South Center Press, pp. 263–296.

Hopenhayn, M. (2002) 'Old and New Forms of Citizenship', *CEPAL Review*, 73: 115–126.

Jones, G. A. (1998) 'Resistance and the Rule of Law in Mexico', *Development and Change*, 26: 499–523.

Jones, G. A. (forthcoming) *Under De-construction: Contesting Modernity and Identity in the Historic Centres of Latin America*. Discussion Paper. London: Institute of Latin American Studies.

Jones, G. A. and Varley, A. (1999) 'Conservation and Gentrification in the Developing World: Recapturing the City Centre', *Environment and Planning A*, 31: 1547–1566.

Kaztman, R. (1997) 'Marginality and Social Integration in Uruguay', *CEPAL Review*, 62: 93–119.

Lara, M. P. (1998) 'The Frail Emergence of Mexico's Democracy: Conquering Public Space', *Thesis Eleven*, 53: 65–78.

Lewis, C. and Pile, S. (1996) 'Women, Body, Space: Rio Carnival and the Politics of Performance', *Gender, Place and Culture*, 3: 27–44.

Lucero, J. A. (2001) 'High Anxiety in the Andes: Crisis and Contention in Ecuador', *Journal of Democracy*, 12 (2): 59–73.

Mainwaring, S. and Scully, T. (eds) (1995) *Building Democratic Institutions: Party Systems in Latin America*. Stanford, CA: Stanford University Press.

Marshall, T. H. (1998) 'Citizenship and Social Class', in G. Shafir (ed.), *The Citizenship Debates*. Minneapolis: University of Minnesota Press.

Massey, D. (1999) 'Spaces of Politics', in D. Massey, J. Allen and P. Sarre (eds), *Human Geography Today*. Cambridge: Polity Press, pp. 279–294.

Melucci, A. (1996) *Challenging Codes: Collective Action in the Information Age*. Cambridge: Cambridge University Press.

Molyneux, M. (2002) 'Gender and the Silences of Social Capital: Lessons from Latin America', *Development and Change*, 33 (2): 167–188.

Mouffe, C. (1993) *The Return of the Political*. London: Verso.

Nolin Hanlon, C. and Shankar, F. (2000) 'Gendered Spaces of Terror and Assault: The Testimonio of REMHI and the Commission for Historical Clarification in Guatemala', *Gender, Place and Culture*, 7: 265–286.

Ocampo, J. A. (1998) 'Income Distribution, Poverty and Social Expenditure in Latin America', *CEPAL Review*, 65: 7–14.

O'Donnell, G. (1999) 'Polyarchies and the (Un)rule of Law in Latin America: A Partial Conclusion', in J. E. Méndez, G. O'Donnell and P. S. Pinheiro (eds), *The (Un) Rule of Law and the Underprivileged in Latin America*. Notre Dame, IN: University of Notre Dame Press, pp. 303–337.

Pearce, J. (1997) 'Civil Society, the Market and Democracy in Latin America', *Democratization*, 4: 57–83.

Pereira Leite, M. (2000) 'Entre o individualismo e a solidariedade: dilemmas da politica e da cidadania no Rio de Janeiro', *Revista Brasileira de Ciencias Sociais*, 44: 73–90.

Philip, G. (1999) 'The Dilemmas of Good Governance: A Latin American Perspective', *Government and Opposition*, 34 (2): 226–242.

Pinheiro, P. S. (1996) 'Democracy without Citizenship?', *NACLA Report on the Americas*, 30: 1–23.

Portes, A. (1998) 'Social Capital: Its Origins and Applications in Modern Sociology', *Annual Review of Sociology*, 24: 1–24.

Putnam, R. D. (1995) 'Bowling Alone: America's Declining Social Capital', *Journal of Democracy*, 6 (1): 65–78.

Radcliffe, S. (1993) 'Women's Place/El Lugar de Mujeres: Latin America and the Politics of Gender Identity', in M. Keith and S. Pile (eds), *Place and the Politics of Identity*. London: Routledge, pp. 102–116.

Roberts, B. (1996) 'The Social Context of Citizenship in Latin America', *International Journal of Urban and Regional Research*, 20: 38–65.

Roberts, K. (1997) 'Beyond Romanticism: Social Movements and the Study of Political Change in Latin America', *Latin American Research Review*, 32 (2): 137–151.

Robinson, W. I. (1999) 'Latin America in the Age of Inequality: Confronting the New "Utopia"', *International Studies Review*, 1 (3): 41–68.

Rosenthal, A. (2000) 'Spectacle, Fear and Protest: A Guide to the History of Urban Public Space in Latin America', *Social Science History*, 24 (1): 33–73.

Routledge, P. (1998) 'Going Globile: Spatiality, Embodiment, and Mediation in the Zapatista Insurgency', in G. Ó Tuathail and S. Dalby (eds), *Rethinking Geopolitics*. London: Routledge, pp. 240–259.

Rubio, M. (1997) 'Perverse Social Capital: Some Evidence from Colombia', *Journal of Economic Issues*, 31: 805–816.

Scarpaci, J. and Frazer, L. (1993) 'State Terror: Ideology, Protest and the Gendering of Landscapes', *Progress in Human Geography*, 17: 1–21.

Schild, V. (1998) 'New Subjects of Rights? Women's Movements and the Construction of Citizenship in the "New Democracies"', in S. Alvarez, E. Dagnino and A. Escobar (eds), *Cultures of Politics, Politics of Cultures: Re-visioning Latin American Social Movements*. Boulder, CO: Westview Press, pp. 93–117.

Schönwälder, G. (1997) 'New Democratic Spaces at the Grassroots? Popular Participation in Latin American Local Governments', *Development and Change*, 28: 753–770.

Sen, A. (1999) 'Democracy as a Universal Value', *Journal of Democracy*, 10 (3): 3–17.

Sennett, R. (1977) *The Fall of Public Man*. Harmondsworth: Penguin.

Shefner, J. (2001) 'From Popular Movements to Public Policy: Outcomes of Democratization in Guadalajara', *Out of the Shadows: Political Action and the Informal Economy*. Paper presented at conference, Univerisity of Princeton, November 15–17.

Slater, D. (2002) 'Other Domains of Democratic Theory: Space, Power, and the Politics of Democratisation', *Environment and Planning D: Society and Space*, 20: 255–276.

de Sousa Santos, B. (1998) 'Participatory Budgeting in Porto Alegre: Toward a Redistributive Democracy', *Politics and Society*, 26: 461–510.

Stephen, L. (2001) 'Gender, Citizenship and the Politics of Identity', *Latin American Perspectives*, 28 (6): 54–69.

Streicker, J. (1997) 'Spatial Reconfigurations, Imagined Geographies, and Social Conflicts in Cartagena, Colombia', *Cultural Anthropology*, 12 (1): 109–128.

Tedesco, L. (2000) 'La ñata contra el vidrio: Urban Violence and Democratic Governability in Argentina', *Bulletin of Latin American Research*, 19: 527–545.

Tendler, J. (1996) *Good Government in the Tropics*. Baltimore, MD: Johns Hopkins University Press.

Waisbord, S. R. (1995) 'Farewell to Public Spaces? Electoral Campaigns and Street Spectacle in Argentina', *Studies in Latin American Popular Culture*, 15: 279–300.

Walzer, M. (1986) 'Pleasures and Costs of Urbanity', *Dissent*, 33: 470–475.

Young, I. M. (1996) 'Communication and the Other: Beyond Deliberative Democracy', in S. Benhabib (ed.), *Democracy and Difference*. Princeton, NJ: Princeton University Press, pp. 120–135.

Young, I. M. (2001) 'Activist Challenges to Deliberative Democracy', *Political Theory*, 29: 670–690.

Zaluar, A. (2001) 'Violence in Rio de Janeiro: Styles of Leisure, Drug Use and Trafficking', *International Social Science Journal*, 53: 369–378.

10 Media, Democracy and Representation: Disembodying the Public

Clive Barnett

In this chapter, I want to make a case for the continuing importance of the concept of the public sphere for helping us to understand the role that media and communications play in constituting the meanings and practices of democracy. In the first section, after introducing Jürgen Habermas's influential account in *The Structural Transformation of the Public Sphere* (1989), I will focus upon the way in which the relationships between media, communication, and representation should be conceptualized. Having established that the public sphere can be understood as referring to various mediated spaces of cultural and political representation, the second section considers the ways of thinking about the geographies of the public sphere that follow from this. The final section illustrates how media publics connect the distinctive rationalities of everyday cultural practices and strategic political action. The elaboration of the concept of the public sphere will be supplemented by examples drawn from research on South African media reform in the 1990s, which illustrate the practical working through of normative assumptions about media, culture, and democracy in a context shaped by contradictory imperatives of national development, democratization, and globalization.

Representing: The Public Sphere

In *The Structural Transformation of the Public Sphere*,[1] Habermas (1989) argues that the historical emergence of an infrastructure of protected public discussion marks a key moment in the transformation, not just of *who* holds political power (the people, as enshrined in the principle of popular sovereignty). It also fundamentally alters the nature of political power itself by transforming *how* power is exercised. Political domination is subordinated to democratic scrutiny by virtue of the accessibility of information to the public, guaranteed by effective rights of free speech, association, and assembly. These enable the active participation of individuals in public discussion and debate. In this way, the exercise of power passes through institutionalized mediums of public deliberation, giving publicly agreed norms a practical efficacy over the actions of economic organizations and political institutions.

The public sphere has a very precise meaning in *The Structural Transformation of the Public Sphere*. It is one element in a four-way division of the social field. The patriarchal family and the market economy both belong to the 'private' realm, while the state is the locus of 'public' authority. The public sphere is defined as an inter-mediating zone between these two realms: the concept refers to the set of practices through which *public opinion* is formed and articulated. In the terms of Habermas's (1984) later social theory, one can divide these four realms into a more complex pattern of cross-cutting relationships: a private realm of communicatively inte-grated *lifeworld* relations (the family); a private realm of *system* relations (the capitalist market economy); a public realm of *system* relations (the state); and a public realm of *lifeworld* relations (the literary and political public spheres). 'Public' and 'private' therefore refer to a distinction between practices governed by an orientation towards universal values (the state and the public sphere) and those governed by particularistic values (the family and the market).

The main limitation of Habermas's original account of the public sphere is his tendency to derive the normative significance of public forms of deliberation and decision-making from very specific historical models (see Calhoun, 1993; Hohendahl, 1979, 1995). In particular, *The Structural Transformation of the Public Sphere* privileges a particular set of cultural institutions that are shaped by inequalities of both class and gender. This specificity is reflected in the implicit cultural theory that identifies particular literary capabilities as the conditions of democratic inclusion. The emphasis on the cultivation of literary competencies is related to the consolidation of capitalist property relations. These define the autonomy of the private (masculine) citizens who engage in public deliberation, and they also facilitate the circulation of a politically free, commercialized, compet-itive market for information and opinion, in the form of books, newspapers, and pamphlets. This cultural theory of democratic competence underwrites the tragic narrative of the decline of the public sphere in the second part of *The Structural Transformation of the Public Sphere*, in which the gendered sub-text of Habermas's account becomes evident. The primary cause of the transformation of the public sphere in the twentieth century is the interweaving of the public and private realms, as the modern state progressively took on responsibilities for the repro-duction of the social relations of commodity production and exchange. This is associated with a process in which both the content of politics and the modes of public communication are presented as becoming progressively feminized, not least by the commodification of public communication with the rise of electronic mass media of radio and television (see Lacey, 1996; McLaughlin, 1993). In turn, the critical function of the public sphere is progressively eroded as the media of public debate are transformed into mediums for the expression of particularistic interests rather than the formation of a universally agreed general interest. The increasingly commodified, as distinct from merely commercialized, mass media become arenas for winning mass approval.

It is when public opinion becomes a mere representation, embodied in intermediaries like opinion polls, pollsters, and experts, rather than being formed through active citizen participation in undistorted communication, that the critical role of the public sphere is undermined. In this model, democratic participation is equated with involvement in rational forms of communication that are oriented to universality, requiring the exchange of ideas between subjects who should properly be indifferent to their own particularistic interests and embodied identities. Accordingly, any attempt to re-shape identities through public action is considered a symptom of a back-sliding towards symbolic forms of so-called representative publicity, in which centralized power is displayed before a passive public. Habermas's narrative of the re-feudalization of the public sphere is therefore characterized by a deep *distrust of representation* (Peters 1993). This is most strongly signalled in a particular conceptualization of print-based cultural practices. The textuality of communicative action is the source of the founding ambivalence of Habermas's original account of the media and the public sphere (Lee, 1992; Saccamano, 1991). On the one hand, the classical liberal public sphere depends upon various print media (pamphlets, newspapers, and novels). However, these mediums are understood merely as conduits for the transmission of information between locales. Habermas explicitly subordinates the disseminating force of writing to the bounded continuities of idealized conversation (Habermas, 1989: 42). The spatial and temporal extension enabled by print media is not therefore allowed to disrupt a model of consensus formation through undistorted dialogue among an essentially homogenous reading public.

Habermas's conceptual reduction of print to an idealized norm of conversation drastically narrows the modes of communication through which the problematization of the 'public interest' and the delineation of 'politics' are allowed to take place. Spectacle, display, and other non-verbal forms of communication are understood to reduce citizens to passive spectators, rather than active participants in deliberation. One line of critical elaboration of Habermas's theory has therefore focused upon positively affirming the role that rhetoric, passion, spectacle, and other modes of affective communication play in public life (Calhoun, 1997; Keane, 1984; Young, 1993, 1997). This conceptual critique itself reflects a shift in the repertoires of political action pursued by a diverse range of social movements (Young, 2001). If this critical work broadens the range of the legitimate forms of democratic communication, it simultaneously acknowledges that the identities of participants are formed and transformed through public communicative practice. And this suggests a revised understanding of the nature of acts of representation. The struggles for representation by a variety of subaltern political subjects replaces an expressive understanding of representation, understood as a means of restoring to presence an already formed identity, with a more strongly constructive, transformative sense of representation. There is no transparent relation of representation because representation is a supplementary process which reveals that identities, interests, or the will are always already non-identical, hence the need for supplementation by an act of representation (Laclau, 1993).

Acknowledging the irreducibility of representation[2] in democratic theory suggests that the concept of 'public' is not best understood as a synonym for a social totality or a collective actor, and nor should it be immediately understood as referring to particular public spaces of bounded social interaction. Rather, the sense of the public that can be gleaned from reading Habermas's original account against the grain refers to a set of processes that exist in relation to a temporally and spatially distanciated network of circulating texts, images, and symbolic acts (see Warner, 2002). The exemplar of this sort of *strung-out, open-ended* public would be those publics constituted by the indeterminate address of various forms of electronic mass media (Scannell, 2000). This understanding is implicit in Habermas's original discussion of a public constituted by circulating print media, but his account needs to be freed from the conceptual containment of writing within a closed circuit of dialogue. Lifting this restriction points towards the conceptual and practical relevance for democratic theory of Claude Lefort's account of the imaginary institutionalization of popular sovereignty in democracy (1988: 9–20). According to this understanding, democracy installs 'the people' as the highest sovereign authority, but in so far as the exercise of this sovereignty is necessarily *staged* through the representational medium of public debate, the public is always already dispersed and constitutively pluralized. This post-foundational understanding of democracy depends on abandoning the normative presumption that the public refers to a self-identical collective subject that could be made present in a space of assembly. Rather, the public is always non-identical with itself (Lefort, 1986: 273–306).[3] It follows that democratic representation is properly thought of as performative, that is to say, as a set of reiterative acts that bring into existence the identities that they appear to be merely re-presenting (Derrida, 1986; Honig, 1991). This is not the equivalent of saying that the public is a mere fiction. This rejectionist position depends on an all-or-nothing understanding of representation, underwritten by a singularly undemocratic logic of identity and authentic presence (Dostal, 1994). Rather, affirming the irreducibility of representation rests on the acknowledgement that democracy depends, at a minimum, on maintaining the right to question the legitimacy, authority, and accountability of unavoidable claims to speak on behalf of absent others.

This sense of the performative character of democratic representation also points to the crucial dimension of temporal non-coincidence of deliberation and decision implied by the concept of public sphere. The distinctive temporal constitution of normative democratic principles is easily elided by the privileging of space as the medium in which multiplicity and plurality can be acknowledged (Massey, 1999). The constitutive relationship between democracy, difference, and conflict finds its clearest practical resolution in the weaving together of overlapping temporal rhythms. The meaning of democratic norms is distilled in the development of mechanisms that institutionalize *contingency*, *reversibility*, and *accountability* into decision-making procedures. This folding of different temporal registers is the means for facilitating legitimate binding decision-making in contexts of non-reconcilable difference, enabling the formation of consensual decision-making freed from a horizon of transcendence but maintaining an orientation to the future (Derrida, 1992; Dunn, 1999).

Acknowledging the temporal dimensions of democratic norms leads to the conclusion that the public sphere should not be thought of as having a definitive, once-and-for-all material form, whether this is understood to be properly embodied in spatial archetypes (e.g. coffee-shops, streets corners, or the idealized heterogeneous public spaces of the cosmopolitan metropolis), or specific institutional configurations (e.g. public service broadcasting). Rather, the public sphere refers to certain sorts of *process*, ones that certainly have social, institutional, and organizational conditions of possibility, but which should not be conflated with any particular configuration of these. This process-based understanding helps us to specify the significance of principles of free speech in democratic theory. It is a commonplace that, in modern democracies, public deliberation is carried on by professional communicators, including pundits, politicians, and journalists (Page, 1996). At one level, this can be seen as a necessary result of a division of labour that follows from the numerical size, geographical scale, and functional complexity of modern societies. More abstractly, however, this returns us to Lefort's argument that democratic authority is an essentially empty place only ever occupied temporarily by proxies (1986: 279). This idea points to the constitutive role of both freedom of speech and freedom of the press principles in democratic theory, as well as the irreducibility of the former to the latter. The significance of free speech principles derives in large part from the distinctive temporal relationship between performative *representation* and retrospective *accountability* that characterizes modern democracy. The protection of basic communicative freedoms works, in principle, to prevent the people's representatives from substituting themselves fully for the represented, thereby appropriating the place of power: 'Freedom of public opinion keeps open the possibility that the represented might at any time make their own voices heard' (Manin, 1997: 174). Media of public opinion give institutional form to the relationship between the irreducible movement of representation, the folded temporalities of deliberation and decision-making, and the originary dispersal of the public that defines democratic norms. This leads on to the conclusion that the public sphere should be thought of as 'structurally elsewhere, neither lost nor in need of recovery or rebuilding but defined by its resistance to being made present' (Keenan, 1993: 135). And this might in turn help us to better understand where the analytical focus of a political geography of the public sphere should be directed.

Rethinking the Geographies of the Public Sphere

Where is Public Space?

One of the recurring problems in discussions of Habermas's original theory is the tendency to over-substantialize the public sphere in specific spatial and/or institutional configurations. In geography and urban studies, reference to *The Structural Transformation of the Public Sphere* usually serves as the preliminary to arguments that Habermas's conceptualization needs to be grounded in relation to an analysis of real, material public spaces. The assumption is that Habermas's original conceptualization of the public sphere depends on a metaphorical understanding of material spaces. Debates around the meanings of the public sphere are consequently registered in geography and related fields by a series of discussions of exemplary forms of public, typically urban, spaces (see Goheen, 1998; Howell, 1993; Light and Smith, 1998; Mitchell, D., 1995; Mitchell, K., 1997; Zukin, 1995). However, the argument that Habermas's public sphere is insufficiently material seems wrong-headed. The problem with *The Structural Transformation of the Public Sphere* is not that it ignores real spaces, but that it conceptually constructs locales of co-presence as the norm for judging the publicness of historically variable practices of social interaction. Furthermore, geographers' determination to translate the public sphere into bounded public urban spaces of co-present social interaction elides Habermas's consistent focus upon the goal-oriented dimensions of public life. It also illustrates a long-standing underestimation of the significance of communications practices in critical human geography (Hillis, 1998).

In contrast to the tendency to romanticize real and material urban public spaces as privileged locations for political action,[4] I want to argue that in a strong and non-metaphorical sense, the media constitute the 'space of politics' implied by the idea of the public sphere (see Dahlgren, 2001: 83–86; see also Bennett and Entmann, 2000). This claim rests on recognizing the extent to which the normative significance of the notion of the public sphere is dependent on a strong understanding of political practice as a form of communicative action. That this in turn implies a geographical dimension to public action is captured by Mitchell's (1995: 115) definition of public space as 'a place within which a political movement can stake out the space that allows it to be seen. In public space, political organizations can represent themselves to a larger population.' It is noteworthy, however, that this definition in no way supports a sharp, evaluative distinction between real public spaces and virtual spaces of the media. There are two reasons for this. First, while city spaces might serve as one model for public communication, it is perfectly reasonable to suppose that the important spaces of political communication so defined are just as likely to be radio and television stations, shopping centres, billboards, or even the envelopes through which public utilities bill their clients (Fiss, 1996; Sunstein, 1995). Secondly, any stark opposition between real material spaces and virtual media spaces does not hold up because it fails to register the extent to which various social movements deploy a range of dramaturgical strategies of protest that construct 'real' spaces as stages through which to mobilize media attention and thereby project their presence through spatially extensive media networks (e.g. Adams, 1996; Barry, 2001; Calhoun, 1989). On both these counts, if public space is defined in relation to opportunities for addressing and interacting with audiences, then there is no reason to assume that such spaces are exemplified by shared locales of either spatial or temporal contiguity, rather than distanciated media and telecommunications networks (Samarajiva and Shields, 1997).

The normative privileging of spaces of contiguity in geography's discussions of public space, as well as reflecting a dominant disciplinary focus upon urban spaces, also derives from a tendency to think of publicness primarily in terms of *sociability* (see Weintraub, 1997). This is an understanding that defines public space primarily as a space for the encounter with pluralistic difference. The emphasis on pluralism as a key element of publicness is certainly important, as is focus upon the cultural dimension of this pluralism. However, this emphasis easily slides towards a purist conception of 'the political' defined in stark opposition to more instrumental, purposive understandings of politics. Defining 'the political' as a space of pluralist sociability puts the normative cart before the pragmatic horse, in so far as the qualities of sociability that many writers often define as the essence of public life actually depend on 'a range of decisions, actions, and policies that cannot emerge from the flow of everyday sociability alone' (Weintraub, 1997: 24). In so far as the meaning of politics implies, at least in part, some consideration of the question 'what is to be done?' (Mulhern, 2000: 169–174), then it follows that 'we cannot exclude the element of strategic action from the concept of the political' (Habermas, 1983: 181). In short, the geography of the public sphere should not be narrowly defined in terms of selected spaces of co-present social interaction, for two related reasons. Normatively, the determination to elide the dependence of *communication* on *means of communications* (Peters, 1999) betrays an idealized image of political action as the expression of authentic identity and clear-cut interests. Empirically, the idealization of 'real' and 'material' spaces closes down a full consideration of the geographical constitution of those strategic practices of needs-interpretation and legitimate decision-making that establish the broader conditions of possibility for social interaction guided by norms of civility and respect.

The Spatiality of Communicative Power

One of the fields where Habermas's emphasis on the linkages between the broad cultural conditions of citizenship and the strategic rationalities of political action has been most usefully explored is in media and communications theory. Habermas's ideas on the public sphere have been used to analyze and evaluate the ways in which the organization of mass communications, and broadcasting in particular, have served as mediums of political citizenship, for inculcating broader habits of sociability, and for expanding the scope of care networks (Murdock, 1993; Scannell, 1989).

Policy discourses and academic theories of the appropriate relationships between media and democratic citizenship have developed in a historical context where broadcasting emerged as a complex of technologies, organizations, and markets that articulated two spatial scales of social activity, the national and the domestic. Broadcasting thus contributed to a process whereby social life became increasingly focused upon the private nuclear family at the same time as the real and imaginary horizons of domestic life were stretched over broader spatial scales through improvements in transport, communications, and mass media (Moores, 1993). The fundamental indeterminacy built into the relationships of power and influence characteristic of spatially extensive communications media has been resolved through a combination of paternalism and protectionism. National institutions determined the sorts of programmes that audiences should and should not have access to, in order to assure the cultivation of appropriate models of citizenship (Collins, 1990). These compromises underlie the elevation of an ideal model of public service broadcasting as the embodiment of public sphere principles (see Collins, 1993). In turn, the paradigm of mass communications, national integration, and liberal democracy has been deployed in a variety of non-Western contexts (Samarajiva and Shields, 1990).

Over the last two decades, stabilized patterns of national regulation of media and communications have been transformed by a set of related processes, including the restructuring of corporate ownership and market control; the development of new communications technologies, such as the Internet and mobile telephony; the increasing convergence of computing, telecommunications, and media; and the reorganization of the scales at which regulatory and policy decisions are made. The dynamic behind these processes is the drive to produce new material and institutional infrastructures for the extension of capital accumulation over larger spatial scales at accelerated pace. Satellite television, video, the Internet, the Walkman, and mobile telephony are all cultural technologies that have been institutionalized in a round of policy-making and corporate restructuring that has been motivated by an explicit aim to re-scale the stable national regimes of policy and regulation that have historically characterized broadcast radio and television cultures. This process has involved a reorientation of the discourses linking media and citizenship. Champions of globalization and the information revolution deploy an understanding of citizenship that focuses upon the expanded choices of media commodities available to citizens as consumers. In contrast, there is a strongly pessimistic strain of policy analysis that sees the increasing accessibility to new media forms made available through globalization as heralding the end of effective national policy regulation in the public interest (e.g. Price, 1995; Tracey, 1998).

At one level, these debates turn on different understandings of the appropriate scale at which media, citizenship, and political power should be connected up (see Morley and Robins, 1995; Schlesinger, 1993, 1997). From another perspective, however, they turn more fundamentally on different understandings of the *spatiality of scale* (Low, 1997). Nicholas Garnham provides the clearest application of the theory of the public sphere to the analysis of media globalization and its implications for cultural citizenship. Garnham's interpretation of globalization is premised on the assumption that the territorial scope of political power must be matched by the territorial scope of a singular universal media public. The public sphere concept, he argues, necessarily implies a strong concept of universality, understood in a procedural sense as a minimum set of shared discursive rules necessary for democratic communication (see Garnham, 2000). On these grounds, globalization is seen as leading to a disempowering fragmentation of the public sphere:

> [T]he problem is to construct systems of democratic accountability integrated with media systems of matching scale that occupy the same social space as that over which economic and political decisions will make an impact. If the impact is universal, then both the political and media systems must be universal. In this sense a series of autonomous public spheres is not sufficient. There must be a single public sphere, even if we might want to conceive of it as made up of a series of subsidiary public spheres, each organized around its own public political sphere, media system, and set of forms and interests. (Garnham, 1993: 264)

From the assumption that democratic citizenship requires a singular and universal public sphere coterminous with the territorial scale at which effective political power is exercised, Garnham deduces that 'the process of cultural globalization is increasingly de-linking cultural production and consumption from a concrete polity and thus a realizable politics' (Garnham, 1997: 70). Consequently, globalization not only disconnects media systems from the hollowed-out nation-state, but, in so doing, it also generates a feeling of powerlessness, expressed in the rise of identity politics.

Garnham's evaluative opposition between the ideal of a universal and singular public sphere versus pluralistic fragmentation triggered by globalization depends upon an unquestioned assumption that political power is naturally territorialized. In order to avoid the rather glum prognosis that this analysis presents, I want to follow John Keane's (1995: 8) suggestion that the public sphere is better understood as 'a complex mosaic of differently sized, overlapping, and interconnected public spheres that force us radically to revise our understanding of public life and its "partner" terms such as public opinion, the public good, and the public/private distinction'. In contrast to the assumption that political power is always exercised within a territorialized power-container of one scale or another, Keane argues that the conceptual relationships between media and democracy should be based on a networked conception of political power. He suggests that the power of large-scale organizations, like states and corporations, depends on 'complex, molecular networks of everyday power relations' (Keane, 1991: 146). From this capillary perspective on power, conceptualizations that prioritize unified and territorially bounded media publics are ill-suited to assessing the progressive potential of contemporary transformations in the spatial organization of media and communications, because they underestimate the possibility for a multiplicity of networked spaces of communicative practice to induce changes in organizations and political institutions. It follows that the key analytical question is not whether effective democratic media publics can be constituted at the same global level to match the jump of scale by capital and by administrative and regulatory authorities, but rather whether (and how) actors embedded at different territorial scales are able to mobilize support and resources through spatially extensive networks of engagement (Cox, 1998).

Media, Movements and the Politics of Scale

South African media reform in the 1990s can be used to illustrate this reformu-lated understanding of the relationships between media, politics, and the networked spatiality of scale. Formal democratization has been associated with an opening up of a previously tightly controlled media system to international investment, a diversification of radio and television outlets, increased levels of competition and commercialization, and the heightened commodification of audiences (Barnett, 1999). At the same time, as part of a broader emergent culture of transparency and public accountability, an infrastructure of independent media regulation has been established. Taken together, these developments have encouraged a shift in practices of journalism, changes in source strategies, and new norms of what counts as 'newsworthy'. In certain circumstances, these structural and organizational shifts have opened up new opportunities for locally embedded social movements to mobilize media attention as a means of applying pressure on local, provincial, and national political elites. For example, environmental activists in Durban have been able to project local protest against industrial pollution upwards to a national scale of radio and television news by gaining extensive coverage in Durban-based newspapers (Barnett, 2003b). There has been a significant increase in the number of stories on 'brown' environmental-justice issues during the 1990s. Furthermore, these stories are increasingly framed to represent the legitimacy of local community mobilization, and which, since 2001 especially, ascribe significant policy changes to this sort of political action. And the relative success of this example of activism has been in part facil-itated by an extensive network of support, maintained through routine Internet and web-based communication with US-based organizations. These networks provide access to various resources, including scientific expertise, discursive frames, and new repertoires of protest, all of which have been critical to the media-oriented strategies of Durban-based activists. In this case, media and com-munications restructuring has therefore opened up opportunities for new forms of political action. These changes are indicative of a genuine transformation of the scope and effectiveness of the South African public sphere in providing critical public scrutiny of government and corporate power in a context of the formal institutionalization of a liberal representative democratic settlement.

In introducing this example, I do not want to idealize the progressive polit-ical potential of 'new media'. Rather, I want to suggest two things. First, this case illustrates that the political significance of communications media is not technologically determined, but in large part depends upon the capacity of social actors for mobilization, organization, and self-representation (Scott and Street, 2000). Secondly, the ability of activists to pursue a new form of political action *through* media spaces and communications networks has been dependent on the sorts of politics that have been going on *around* the media in South Africa in this period. The example of environmental activism in Durban illustrates that the material geographies of the public sphere are made up of flows of information, opinion, and ideas that articulate different institutional sites (the media, the home, the workplace, the state) and spatial scales. There is nothing virtual about the publics bought into existence through these networks, whether their medium is the Internet, newspapers, talk-radio or television, if 'virtual' is meant to imply that they are somehow immaterial. These networks are material in a double sense: they are embedded in a tangible infrastructure of institutions, organizations, technologies, and social configurations that are every bit as produced as roads, railways, and buildings (e.g. Streeter, 1996); and they are material in the sense of being effective in shaping opinions, decisions and outcomes. Understanding the *production of communicative spaces* (Barnett, 2003a) therefore requires us to consider both the geographies of collective action constituted through media and communications networks, and the sorts of organization, mobilization, and interest-group representation that emerge around issues of media access, cross-media ownership, privacy, universal service and so on (see McChesney, 1993; Schiller, 1999).

By insisting upon the importance of analyzing broader patterns of political organization, I also want to counter a tendency to present new media and communications technologies as providing solutions to unfortunate empirical obstacles of mediation, distanciation, and representation. Images of new communications technologies inaugurating a new age of direct democracy should be treated with deep suspicion. Presenting the Internet, for example, as a technological fix that enables instantaneous plebiscites on any number of topics betrays an atomistic model of democracy, understood as the expression of privately formed preferences (Elster, 1986). By effacing the public formation of interests and identities, it is an understanding that elides a set of crucial questions about the social determination of preferences (see Sunstein, 1992).

Again, the South African case provides an example of why the rhetoric of technologically induced direct democracy fails to capture the full complexity of the relationships between media, democracy, and the formation of citizenship. Historically, South African press, publishing, and broadcasting has not provided a common space of shared public communication. These media have been used to reproduce notions of separate and distinct populations, with their own separate cultures, belonging in separate geographical areas. Consequently, South African citizens have starkly unequal capacities to express their cultural and political preferences through individualized, commodified forms of media provision. Since the end of Apartheid in 1994, broadcasting and telecommunications policies have aimed to foster national integration in an international context of increasing globalization of both sectors (Horwitz, 2001). In seeking to overcome the divisive legacies of apartheid media policies, the role of various collective actors from 'civil society', in arguing for a politically independent and financially viable public service broadcaster, as well as an accountable system of independent media regulation, has been critical in opening up opportunities for the expression of tastes, interests, and identities that would have otherwise been silenced by a shift towards a fully privatized, deregulated and market-led media system. One of the most important impacts of international media and telecommunications policy in the 1990s has been the proliferation of independent communications regulators, set up to oversee newly or soon-to-be privatized and liberalized national communications sectors. In this process, the shaping of independent regulatory authorities has become a new site through which citizen participation can be channelled. In South Africa in the mid-1990s, the politics of independent communications regulation saw significant successes for progressive organizations in embedding procedures for accountability, transparency, and public participation in communications policy. This success has been pivotal in the pluralization of media cultures and the democratization of news agendas. However, the ongoing internationalization of South African communications policy has more recently seen the degree of civil participation and democratic accountability curtailed by a prioritization of investment-led regulatory principles. In turn, there is an emerging network of Southern African media activism, sharing information and expertise, and engaging in multiple policy contexts (Barnett, 2002a). This internationalization of media reform movements underscores the point that the success of campaigns for the continuing democratization of mass media *within* South Africa will be shaped by the capacity of nationally embedded actors to draw upon networks of political support and institutional resources that stretch beyond the confines of the nation-state.

Cultural Public Spheres and Mediated Deliberation

My argument to this point has been that a process-based understanding of the public sphere, understood as an institutionalized space of representation, directs attention to the role of media and communications practices in mediating the possibilities for an expansive politics of scale. In this final section, I want to argue that, as well as being important for facilitating certain sorts of explicitly political action in the strong sense, the public sphere concept also directs attention to the ways in which the media articulate the distinctive rhythms of everyday cultural practice with the imperatives of strategic political action. In their critical revision of Habermas's original account of the public sphere, Negt and Kluge (1993) develop an understanding of the public sphere as a diffuse medium of cultural democratization. They extend the notion of the public sphere to include the relationships which constitute the very conditions of possibility for social and individual experience:

Public sphere refers to certain institutions, establishment, activities (e.g. public power, press, public opinion, audience, publicity work, streets, and squares); but at the same time it is also a general social horizon of experience, in which what is really and supposedly relevant for all members of a society is summarized. In this sense the public sphere is a matter of a few professionals (e.g. politicians, editors, officials or federations); on the other it is something that has to do with everybody and which is only realized in the heads of people, a dimension of their consciousness. (Negt and Kluge, 1993: 1–2)

This emphasis upon multiple publics is linked to a pluralization of the modes of public communication through which interests, needs, identities, and desires can be legitimately articulated. This pluralization requires that Habermas's focus on the literary formation of democratic competencies be recast in terms of a cultural public sphere (McGuigan, 1998). This term refers to a wider array of affective communicative and expressive practices of popular culture, in contrast to narrowly cognitive and rational understandings of deliberation, and to the institutional and social determination of the distribution of the cultural capabilities.

The notion of the cultural public sphere implies a broader understanding of how the cultural articulates with the political, without reducing the latter to the former. A fruitful way of understanding this relationship is provided by Nancy Fraser's analytical distinction between *weak* and *strong* public spheres. Weak publics refer to those activities 'whose deliberative practice consists exclusively in opinion-formation and does not encompass decision-making'. Strong publics, more directly connected to institutionalized power, are those activities in which 'discourse encompasses both opinion-formation and decision-making' (Fraser, 1997: 90). In his recent work, Habermas adopts this distinction, and accords considerable importance to weak publics, understood as a 'wild complex' of informal processes of opinion formation (Habermas, 1996: 307–308). Habermas's use of the weak/strong distinction is related to an abandonment of his earlier pessimistic 'siege' model of relationships between lifeworld and system, in which social movements were understood as being in an essentially defensive posture against the 'colonization of the lifeworld' (Habermas, 1984). At a conceptual level, Habermas has become notably more optimistic about the possibilities for the effective democratic oversight of administrative and economic power. His revised conception of radical democracy is premised upon a decentred image of society:

The public sphere cannot be conceived as an institution and certainly not as an organization. It is not even a framework of norms with differentiated competencies and roles, membership regulations, and so on. Just as little does it represent a system; although it permits one to draw internal boundaries, outwardly it is characterized by open, permeable, and shifting horizons. The public sphere can best be described as a network for communicating information and points of view (i.e. opinions expressing affirmative or negative attitudes); the streams of communication are, in the process, filtered and synthesized in such a way that they coalesce into bundles of topically specified public opinions. (Habermas, 1996: 360)

Thus, in his recent work, Habermas develops a conception of the public sphere as a network of decentred 'streams of public communication', without any a priori restriction placed on their precise form or medium. This shift is reflected in the argument that communicative action in weak publics articulates with centres of decision-making through a series of 'sluices', which provide effective critical leverage over economic and political systems.

Although Habermas's earlier suspicion of electronic media has been significantly revised (Habermas, 1992), there is still a degree of ambivalence in his approach to this topic. He is still prone towards interpreting the detachment of public communication from the concrete presence of an audience as leading to a problematic differentiation 'among organisers, speakers, and hearers; arenas and galleries, stage and viewing space' (Habermas, 1996: 363). The tone of this description is indicative of a residual attachment to a dichotomy between active participation and passive spectatorship. I want to make two related points in this respect, in order to more strongly defend the irreducible role of mediated communicative action in shaping the possibilities for the sort of decentred and subject-less streams of communicative power that Habermas identifies as being central to any radical democratic vision.

First, the dispersal of the public through electronic mass media (and now digital communications technologies) underscores the need to rethink the relationships between abstraction and embodiment in political theory. In Habermas's original account of the public sphere, abstraction and universality are opposed to embodiment and particularity, as indicated by his conceptual containment of writing. Representation is hence restricted to the representation of ideas. Following Michael Warner, it is more appropriate to see pluralized public spheres as being characterized by communicative practices that involve a two-way movement between self-abstraction and self-realization. Contemporary forms of identity politics simultaneously affirm particular embodied identities while reaching out to broader identifications in an orientation towards universality without transcendence (Warner, 1993: 252). The decline of the norm of a single public thus disembodies *the* public, understood as a collective subject made present in spaces of assembly, while opening up a cultural politics where the representation of ideas is supplemented by a 'politics of presence', in which the representative value of certain irreducibly embodied identities is acknowledged (Phillips, 1995). The media have become the key site for this sort of cultural action, in both formal politics and in popular culture. The ascendancy of 'audience-democracy' (Manin, 1997: 235), in which the relationship of accountability between representatives and their constituencies becomes centred upon *trust*, suggests that successful political communication has a lot to do with the ability to credibly embody and perform certain sorts of persona (Corner, 2000). This notion can, of course, confirm a self-righteous denunciation of the recidivist dumbing-down of rational political discourse. But it might be given an alternative inflection, as indicative of a fundamental cultural democratization of formal politics that is in large part determined by the most intimate characteristics of distanciated mediums of public communications, through which the norms of the everyday take on a heightened significance in disciplining the conduct of formal public life (Scannell, 1995; Thompson, 2000).

If this first point suggests that rethinking the performative aspects of embodiment among selected actors in public media cultures is an important counterpoint to overly rationalistic accounts of the public sphere, then it still needs to be attached to a sense of the unprecedented enlargement of public participation that modern media cultures facilitate. My second point here, then, is that rather than thinking of the differentiation between 'speakers' and 'hearers', 'stage' and 'viewing space' in terms of an opposition between active participation and passive spectatorship, we should instead consider the media as opening up spaces of *mediated deliberation* (see Thompson, 1995: 125–134). At its simplest, this refers to the re-embedding of mass circulated symbolic materials into contexts of face-to-face dialogical interaction. In this way, modern media and communications technologies vastly expand the range of information, ideas, and opinions made available to larger numbers of ordinary people than ever before. John Dewey (1927) saw this as one of the main contributions of modern communications technologies in expanding the scope and power of public action. However, the concept of mediated deliberation suggests a stronger emphasis upon the necessarily mediated character of any and all deliberative practice. And this points up the extent to which the sort of civil, reciprocal, rule-bound talk often idealized by theorists of deliberative democracy is in fact dependent on the provision of an infrastructure of technologies, institutions, and social and cultural norms that cannot be bought into existence through conversation alone (Schudson, 1997). The notion of mediated deliberation therefore reminds us that the main focus in assessing the social, cultural and political significance of different media practices should remain upon the socio-economic and institutional distribution of the 'capabilities' to engage in the streams of discourse that effectively articulate with centres of decision-making (Garnham, 1999).

I have argued that any analysis of the geographies of the public sphere needs to acknowledge that public participation in spatially extensive and functionally complex societies is necessarily mediated. There are two senses in which this is the case. First, participation in discourses about matters of public importance consists of innumerable practices that revolve around the consumption of books, newspapers, radio, television programmes, pop songs, and other symbolic resources circulated by media and communications industries and institutions. Secondly, these dispersed cultural practices are shaped by, and in turn shape, conflicts between collective actors of various sorts, who speak and act in the name of broader constituencies, and struggle over the framing of the form, content, and scope of 'politics' and 'culture'. These two points underscore the need to avoid an excessively media-centric approach to assessing the relationships between media, democracy and citizenship. Craig Calhoun observes that in modern societies:

> democracy depends on the possibility of a critical public discourse which escapes the limits of face-to-face interaction. This means, in part, finding ways to make the space transcending mass media supportive of public life. It also means developing social arrangements in which local discussions are both possible and able to feed into larger discussions mediated both by technology and by gatherings of representatives. (Calhoun, 1989: 68–69)

In closing, I want to briefly discuss one example of public policy that demonstrates the relevance of using this broad understanding of connections between media, everyday life, and public space as the means of assessing the contribution of media practices to broader processes of formal and informal democratization.

The example is a South African television drama series, *Yizo Yizo*, first broadcast by the national public service broadcaster, the SABC, in 1999, with a second series aired in 2001. The series is one example of a strong commitment that the newly independent SABC has shown since 1994 to innovative and broadly conceived educational uses of its radio and television services. These initiatives use locally produced programming as one element of a very broad strategy of education for citizenship, illustrating the ways in which media can be made supportive of public life by expanding the horizons of normative debate. The series focuses upon the lives of the children, teachers, and parents of a fictional township school. Made in collaboration with the Department of Education, the series was intended to reveal the depth and complexity of the crisis facing South African schools, to encourage a culture of learning and teaching, and to stimulate discussion of key educational issues. *Yizo Yizo* is broadcast at prime time in the evenings, in order to ensure maximum audience exposure among both children and adults, and it rapidly established a large audience, making it the most watched programme on South African TV.

Yizo Yizo has been highly successful in its primary public policy-related objective of opening up the educational crisis in South Africa to broad public debate and interpersonal discussion. The significance of the series lies in making visible a set of issues concerning the conditions of South Africa's school system, opening up to debate sensitive issues such as sexual harassment, rape, gangsterism, and drug abuse. This success illustrates the potential for public broadcasters to use television as a means of localizing the norms of a global audio-visual culture in an attempt to shift the norms of everyday conduct and stimulate public debate. *Yizo Yizo* deploys a range of popular culture styles and genres to address both young and adult audiences. In particular, the series' success depends in no small part on its conscious use of the aesthetic conventions of an increasingly internationalized and commercialized South African popular culture. In its formulation and development, the series acknowledges the existence of youth audiences who have sophisticated cultural literacies. This dimension of the series' success is reflected not simply in the programmes themselves, but also in the extensive multi-media strategy developed by the SABC and Department of Education to support the broader objectives of the series, including a magazine, a radio talk show, and a soundtrack CD featuring local *kwaito* artists. These resources are aimed at fostering public debate around the issues addressed in the series, by providing resources for students, teachers, and parents to engage with the issues raised by the series. This form of educational broadcasting serves as an example of a developmental use of electronic media to expand the capabilities of citizens to exercise effective and substantive communicative freedoms.

Yizo Yizo therefore illustrates the potential for broadcasting to link up the everyday experiences of ordinary people with broader political debates, by facilitating a set of mediated discussions in homes, classrooms, playgrounds, as well as on radio, television, and in newspapers. Again, I do not want to idealize this example (see Barnett, 2002b). The ability to participate in the sort of extended, mediated public debate stimulated by *Yizo Yizo* remains socially uneven, shaped as it is by inequalities of access to social and material resources which are the conditions of participation in informed debates about both public policy and popular culture. The series also raises a set of issues about the role of an increasingly commercialized South African broadcasting system in the commodification of 'black youth markets', and how this process might contribute to the broader segmentation of social groups that will entrench inequalities of access to media technologies and cultural competencies. Nonetheless, *Yizo Yizo* demonstrates the continuing potential for radical participatory media paradigms to flourish by redeploying the norms of global culture for progressive democratic ends. In an era of media abundance, traditional forms of media regulation have been rendered problematic by the spatial restructuring of media markets and new technologies. What the example of *Yizo Yizo* suggests is that supporting citizenship-participation by using locally produced media programming works best when the multiple and increasingly globalized cultural literacies of citizens is acknowledged.

Conclusion

In this chapter, I have suggested that the constitutive relationship between norms of democratic discourse and acts of cultural and political representation requires a conceptualization of the public sphere as a range of institutionalized *processes* of public communication. While I have argued for 'stretching-out' the public in light of a consideration of the importance of media practices in public communication, I have also indicated that the media needs to be decentred from an evaluation of the relationships between the public sphere and the media. Habermas's recent account of the subjectless streams of public communication redirects our attention to the fact that the vitality of any public sphere depends not simply on structures of media organization, ownership, and use, but more broadly depends on the existence of a plurality of modes of social organization and association. This was the theme pursued through the examples of the politics of South African media reform over the last decade. They illustrate three overlapping themes: that media systems facilitate different forms of political action by grassroots organizations; that the sorts of political opportunity opened up by new media and communications technologies are shaped by a politics of regulation that goes on *around* media; and that the *public* significance of media and communications practices has as much to do with everyday cultural practices and norms as it does with more obvious forms of political action. A critical geographical analysis of contemporary public space therefore needs to consider how media and communications practices serve as both weak and strong publics. This requires moving beyond a disciplinary privilege accorded to 'real spaces' as the paradigm of public space, and analyzing communications media as mediums for the spatial and temporal articulation of different forms of social practice, oriented towards different rationalities, and stretched out across different territorial scales. Increasingly, media publics articulate cultural norms that are no longer contained within national regulatory spaces. And the dimensions of media publics are shaped by a politics of collective action that itself reveals media and communications practices to be central to a networked politics of scale that connects up territorially embedded interests with spatially extensive networks of resources.

In short, the notion of the public sphere is a crucial conceptual and evaluative resource for assessing the role of media and communications practices in sustaining democratic social practices. But the continuing salience of this concept depends upon revising some cherished assumptions that often shape academic discourses about media and democracy. These include assumptions about what counts as rationality, what counts as politics, and about what counts as proper conduct in public life, as well as assumptions about the public value of popular media cultures. In a sense, then, it is academic understandings of the relationships between media, culture, and citizenship that are most in need of being democratized.

Notes

Research reported in this chapter was supported by the Leverhulme Trust.

1 Originally published in German in 1962.
2 Throughout this chapter, I use 'representation' in a deliberately multifaceted sense, although always referring to the notion that representation is the act of making present in some sense something that remains literally absent (Pitkin, 1967). See Barnett (2003a: Chapter 1) for further discussion of this paradoxical sense of representation and its significance for thinking about the spatiality of democratic publicity.
3 There is a strong family resemblance between Lefort's account and Habermas's (1996: 463–490) recent procedural theorization of democratic popular sovereignty.
4 See Keith (1996) and Staeheli (1996) for counterpoints to this conflation of political action with particular spaces.

References

Adams, P. (1996) 'Protest and the scale politics of telecommunications', *Political Geography*, 5: 419–441.

Barnett, C. (1999) 'Broadcasting the rainbow nation', *Antipode*, 31: 274–303.

Barnett, C. (2002a) 'Media, democratization and scale', in K. Tomaselli and H. Dunn (eds), *Media, Democracy and Reconstruction*. Colorado Springs: International Academic Publishers.

Barnett, C. (2002b) 'More than just TV': Educational broadcasting and popular culture in South Africa', in C. von Feilitzen and U. Carlsson (eds), *Children, Young People, and Media Globalisation*. Göteborg: UNESCO Clearinghouse on Children, Youth and Media, NORDICOM, pp. 95–110.

Barnett, C. (2003a) *Culture and Democracy: Media, Space and Representation*. Edinburgh: Edinburgh University Press.

Barnett, C. (2003b) 'Media transformation and new practices of citizenship: The example of environmental activism in post-apartheid Durban', *Transformation: Critical Perspectives on Southern Africa*, 52: 1–24.

Barry, A. (2001) *Political Machines: Governing a Technological Society*. London: Athlone.

Bennett, W. L. and Entmann, R. M. (eds) (2000) *Mediated Politics: Communication in the Future of Democracy*. Cambridge: Cambridge University Press.

Calhoun, C. (1989) 'Tiananmen, television and the public sphere: Internationalization of culture and the Beijing Spring of 1989', *Public Culture*, 2 (1): 54–71.

Calhoun, C. (1993) 'Civil society and the public sphere', *Public Culture*, 5: 267–280.

Calhoun, C. (1997) 'Plurality, promises, and public spaces', in C. Calhoun and J. McGowan (eds), *Hannah Arendt and the Meaning of Politics*. Minneapolis: University of Minnesota Press, pp. 232–259.

Collins, R. (1990) *Culture, Communication and National Identity*. Toronto: University of Toronto Press.

Collins, R. (1993) 'Public service and the market ten years on: Reflections on critical theory and the debate on broadcasting policy in the UK', *Screen*, 34: 243–259.

Corner, J. (2000) 'Mediated persona and political culture', *European Journal of Cultural Studies*, 3: 386–402.

Cox, K. (1998) 'Spaces of dependence, spaces of engagement and the politics of scale', *Political Geography*, 17: 1–23.

Dahlgren, P. (2001) 'The transformation of democracy', in B. Axford and R. Huggins (eds) *New Media and Politics*. London: Sage, pp. 64–88.

Derrida, J. (1986) 'Declarations of independence', *New Political Science*, 15: 7–16.

Derrida, J. (1992) *The Other Heading*. Bloomington: Indiana University Press.

Dewey, J. (1927) *The Public and its Problems*. Athens, OH: Ohio University Press.

Dostal, R. J. (1994) 'The public and the people: Heidegger's illiberal politics', *Review of Metaphysics*, 47: 517–555.

Dunn, J. (1999) 'Situating democratic political accountability', in A. Przeworski, S. C. Stokes and B. Manin (eds), *Democracy, Accountability and Representation*. Cambridge: Cambridge University Press, pp. 329–34.

Elster, J. (1986) 'The market and the forum: Three varieties of political theory', in J. Elster and A. Hylland (eds), *Foundations of Social Choice Theory*. Cambridge: Cambridge University Press, pp. 103–132.

Fiss, O. (1996) *The Irony of Free Speech*. Cambridge, MA: Harvard University Press.

Fraser, N. (1997) *Justice Interruptus*. London: Routledge.

Garnham, N. (1993) 'The mass media, cultural identity, and the public sphere in the modern world', *Public Culture*, 5: 251–265.

Garnham, N. (1997) 'Political economy and the practice of cultural studies', in M. Ferguson and P. Golding (eds), *Cultural Studies in Question*. London: Sage, pp. 56–73.

Garnham, N. (1999) 'Amartya Sen's "capabilities" approach to the evaluation of welfare: Its application to communications', in A. Calabrese and J.-C. Burgelman (eds), *Communication, Citizenship, and Social Policy*. Lanham, MD: Rowman and Littlefield, pp. 113–124.

Garnham, N. (2000) *Emancipation, the Media, and Modernity*. Oxford: Oxford University Press.

Goheen, P. (1998) 'Public space and the geography of the modern city', *Progress in Human Geography*, 22: 479–496.

Habermas, J. (1983) 'Hannah Arendt: On the concept of power', in *Philosophical-Political Profiles*. Cambridge, MA: MIT Press.

Habermas, J. (1984) *The Theory of Communicative Action* (Volume 1). Cambridge: Polity Press.

Habermas, J. (1989) *The Structural Transformation of the Public Sphere*. Cambridge, MA: MIT Press.

Habermas, J. (1992) 'Further reflections on the public sphere', in C. Calhoun (ed.), *Habermas and the Public Sphere*. Cambridge, MA; MIT Press, pp. 421–461.

Habermas, J. (1996) *Between Facts and Norms*. Cambridge: Polity Press.

Hillis, K. (1998) 'On the margins: The invisibility of communications in geography', *Progress in Human Geography*, 22 (4): 543–566.

Hohendahl, P. U. (1979) 'Critical theory, public sphere and culture: Jürgen Habermas and his critics', *New German Critique*, 16: 89–118.

Hohendahl, P. U. (1995) 'Recasting the public sphere', *October*, 73: 27–54.

Honig, B. (1991) 'Declarations of independence: Arendt and Derrida on the problem of founding a republic', *American Political Science Review*, 85: 97–113.

Horwitz, R. (2001) *Communication and Democratic Reform in South Africa*. Cambridge: Cambridge University Press.

Howell, P. (1993) 'Public space and the public sphere', *Environment and Planning D: Society and Space*, 11: 303–322.

Keane, J. (1984) *Public Life and Late Capitalism*. Cambridge: Cambridge University Press.

Keane, J. (1991) *Media and Democracy*. Cambridge: Polity Press.

Keane, J. (1995) 'Structural transformations of the public sphere', *Communication Review*, 1: 1–22.

Keenan, T. (1993) 'Windows: Of vulnerability', in B. Robbins (ed.), *The Phantom Public Sphere*. Minneapolis: University of Minnesota Press, pp. 121–141.

Keith, M. (1996) 'Street sensibility? Negotiating the political by articulating the spatial', in A. Merrifield and E. Swyngedouw (eds), *The Urbanization of Injustice*. London: Lawrence and Wishart, pp. 137–160.

Lacey, K. (1996) *Feminine Frequencies: Gender, German Radio, and the Public Sphere, 1923–1945*. Ann Arbor: University of Michigan Press.

Laclau, E. (1993) 'Power and representation', in M. Poster (ed.), *Politics, Theory, and Contemporary Culture*. New York: Columbia University Press. pp. 277–296.

Lee, B. (1992) 'Textuality, mediation, and public discourse', in C. Calhoun (ed.), *Habermas and the Public Sphere*. Cambridge, MA: MIT Press, pp. 402–418.

Lefort, C. (1986) *The Political Forms of Modern Society*. Cambridge: Polity Press.

Lefort, C. (1988) *Democracy and Political Theory*. Cambridge: Polity Press.

Light, A. and Smith, J. (eds) (1998) *The Production of Public Space*. Lanham, MD: Rowman and Littlefield.

Low, M. M. (1997) 'Representation unbound: Globalization and democracy', in K. Cox (ed.), *Spaces of Globalization*. New York: Guilford Press, pp. 240–280.

Manin, B. (1997) *The Principles of Representative Government*. Cambridge: Cambridge University Press.

Massey, D. (1999) 'Spaces of politics', in D. Massey, J. Allen and P. Sarre (eds), *Human Geography Today*. Cambridge: Polity Press, pp. 279–294.

McChesney, R. (1993) *Telecommuncations, Mass Media, and Democracy*. Oxford: Oxford University Press.

McLaughlin, L. (1993) 'Feminism, the public sphere, media and democracy', *Media Culture and Society*, 15: 599–620.

McGuigan, J. (1998) 'What price the public sphere?', in D. K. Thussu (ed.), *Electronic Empires: Global Media and Local Resistance*. London, Arnold, pp. 91–107.

Mitchell, D. (1995) 'The end of public space? People's park, definitions of the public, and democracy', *Annals of the Association of American Geographers*, 85: 108–133.

Mitchell, K. (1997) 'Conflicting geographies of democracy and the public sphere in Vancouver, B.C.', *Transactions of the Institute of British Geographers*, 22: 162–179.

Moores, S. (1993) 'Television, geography and "mobile privatization", *European Journal of Communication*, 8: 365–379.

Morley, D. and Robbins, K. (1995) *Spaces of Identity: Global Media, Electronic Landscapes and Cultural Boundaries*. London: Routledge.

Mulhern, F. (2000) *Metaculture*. London: Routledge.

Murdock, G. (1993) 'Communications and the constitution of modernity', *Media, Culture and Society*, 15: 521–539.

Negt, O. and Kluge, A. (1993) *Public Sphere and Experience*. Minneapolis: University of Minnesota Press.

Page, B. (1996) *Who Deliberates? Mass Media in Modern Democracy*. Chicago: University of Chicago Press.

Peters, J. D. (1993) 'Distrust of representation: Habermas on the public sphere', *Media, Culture and Society*, 15: 541–571.

Peters, J. D. (1999) *Speaking into the Air*. Chicago: University of Chicago.

Phillips, A. (1995) *The Politics of Presence*. Cambridge: Polity Press.

Pitkin, H. (1967) *The Concept of Representation*. Berkeley: University of California.

Price, M. E. (1995) *Television, the Public Sphere, and National Identity*. Oxford: Oxford University Press.

Saccamano, N. (1991) 'The consolations of ambivalence: Habermas and the public sphere', *Modern Language Notes*, 106: 685–698.

Samarajiva, R. and Shields, P. (1990) 'Integration, telecommunication, and development', *Journal of Communication*, 40 (3): 84–105.

Samarajiva, R. and Shields, P. (1997) 'Telecommunications networks as social space', *Media, Culture and Society*, 19: 535–555.

Scannell, P. (1989) 'Public service broadcasting and modern public life', *Media, Culture and Society*, 11: 135–166.

Scannell, P. (1995) *Radio, Television, and Modern Life*. Oxford: Blackwell.

Scannell, P. (2000) 'For-anyone-as-someone structures', *Media, Culture and Society*, 22: 5–25.

Schiller, D. (1999) 'Social movements in telecommunications', in A. Calabrese and J.-C. Burgelman (eds), *Communication, Citizenship, and Social Policy*. Lanham, MD: Rowman and Littlefield, pp. 137–155.

Schlesinger, P. (1993) 'Wishful thinking: Cultural politics, media, and collective identities in Europe', *Journal of Communication*, 43: 6–17.

Schlesinger, P. (1997) 'From cultural defence to political culture: Media, politics and collective identity in the European Union', *Media, Culture and Society*, 19: 369–391.

Schudson, M. (1997) 'Why conversation is not the soul of democracy', *Critical Studies in Mass Communication*, 14: 297–309.

Scott, A. and Street, J. (2000) 'From media politics to e-protest: The use of popular culture and new media in parties and social movements', *Information, Communication and Society*, 3: 215–240.

Staeheli, L. (1996) 'Publicity, privacy and women's political action', *Environment and Planning D: Society and space*, 14: 601–619.

Streeter, T. (1996) *Selling the Air*. Chicago: University of Chicago Press.

Sunstein, C. R. (1992) 'Preferences and politics', *Philosophy and Public Affairs*, 20: 3–34.

Sunstein, C. R. (1995) *Democracy and the Problem of Free Speech*. New York: Free Press.

Thompson, J. B. (1995) *The Media and Modernity*. Cambridge: Polity Press.

Thompson, J. B. (2000) *Political Scandal: Power and Visibility in the Media Age*. Cambridge: Polity Press.

Tracey, M. (1998) *The Decline and Fall of Public Service Broadcasting*. Oxford: Oxford University Press.

Warner, M. (1993) 'The mass public and the mass subject', in B. Robbins (ed.), *The Phantom Public*. Minneapolis: University of Minnesota Press.

Warner, M. (2002) *Publics and Counterpublics*. New York: Zone Books.

Weintraub, J. (1997) 'The theory and politics of the public/private distinction', in J. Weintraub and K. Kumar (eds), *Public and Private in Thought and Practice*. Chicago: University of Chicago Press, pp. 1–42.

Young, I. M. (1993) 'Justice and communicative democracy', in R. Gottleib (ed.), *Radical Philosophy*. Philadelphia: Temple University Press, pp. 123–143.

Young, I. M. (1997) *Intersecting Voices*. Princeton, NJ: Princeton Univeristy Press.

Young, I. M. (2001) 'Activist challenges to deliberative democracy', *Political Theory*, 29: 670–690.

Zukin, S. (1995) *The Cultures of Cities*. Oxford: Blackwell.

11 Cultures of Democracy: Spaces of Democratic Possibility

Sophie Watson

In 1993 Robert Putnam published his influential book, *Making Democracy Work*, on the importance of building social capital for democracy. Though this work, and his more recent book *Bowling Alone* (2000) have not been central to geographical thinking, it is in my view important for geographers to engage with Putnam's ideas. The reason for this is that Putnam's arguments rely on an implicit version of space and visibility where political association and democratic engagement are examined across space, that is, in different regions, cities and localities of Italy or the USA. What is missing is an exploration of democratic cultures that occur within different or less obvious spatial forms where these might be domestic, interstitial, temporary, or fluid. My argument is that in taking space as a fixed and visible frame for his argument, many more interesting cultural forms of politics get overlooked. New forms of cultural geography can bring a more nuanced approach to Putnam's argument. This is important since many contemporary political accounts and policy approaches rely heavily on Putnam's arguments, or on similar political analysis that is likewise spatially under-theorized.

Putnam on Social Capital, Democracy and Community

In *Making Democracy Work*, through a study of the 20 regional governments in Italy in the 1970s, Putnam establishes a connection between strong networks of civic engagement such as neighbourhood associations, cooperatives and sports clubs, all characterized by intense horizontal interaction, and formal democratic participation. A vertical network no matter how dense cannot sustain social trust and cooperation in the same way (1993: 174). For Putnam these horizontal networks are an essential form of social capital, which foster robust forms of reciprocity, social trust and norms of acceptable behaviour, which are mutually expected, performed and reinforced (1993: 172). The more communities participate actively in mutual exchange displaying their trust of one another, the more other forms of social capital will develop. Thus, in social contexts where group members trust one another, sharing their skills, labour, knowledge, tools, a strong society, economy and state will ensue. In the Italian context, the regions with a large number of active community organizations, such as Emilia-Romagna and Tuscany, were also those where democracy appeared to be working, whereas the 'uncivic' regions like Calabria and Sicily were revealed to have limited engagement in social and cultural association and less effective representative government. The book concludes with a plea for the building of social capital, which Putnam argues, may not be easy, but which is the key to making democracy work.

In *Bowling Alone*, Putnam develops these ideas in the American context. Here, his concern is to explain what he perceives to be the decline in civic engagement and erosion of social capital over the last three or so decades. Using the metaphor of 'bowling alone', referring to a case of kinship felt between two members (one white, one African-American) of a bowling club where one man donated a kidney to the other, Putnam argues that Americans need to reconnect with one another and rebuild social capital. That these two men 'bowled together made all the difference' for Putnam (2000: 28). Developing further the notion of social capital, he draws a distinction between *bridging* (or inclusive) and *bonding* (or exclusive) forms of social capital (2000: 22). The former describes associations like the civil rights associations or youth service groups, and is better for fostering linkage to external assets and information diffusion. The latter characterizes inward-looking groups, such as ethnic fraternal organizations and fashionable country clubs, which tend towards homogeneity and reinforce exclusive identities and are better for underpinning specific reciprocity and mobilizing solidarity. The latter, though good for creating strong group loyalty and for 'getting by', as opposed to 'getting ahead', has the potential for creating strong out-group antagonism (or, in a different discourse, for 'demonizing the other'). At its worst, this is reflected in organizations like the Ku Klux Klan, and at the very least may be associated with high levels of intolerance. This is for Putnam 'the dark side of social capital'. The distinction drawn here has some echoes with Castells' (1997) discussion of new social movements, or identities, as leading to a whole range of heavens or of hells, or more likely heavenly hells.

From a detailed study of Americans' participation in a wide range of arenas, from public politics to clubs and community associations, religious bodies and work-related organizations, Putnam draws the conclusion that following the deep levels of engagement in community life in the first two decades of the century, in the last three decades people have been pulled apart from one another and from their communities. I have focused on Putnam's arguments because they have been both useful and influential, and also because I want to take issue with them in a number of ways. This chapter aims to do several things. First, I want to suggest that though Putnam is analyzing an important phenomenon in his two books, we may need also to look in different places for new cultures of democracy and new forms of association. There may, in other words, be a spatial reordering of community and social capital taking place, which at times may not be obvious, which may be shifting or momentary, and which may even be invisible to all but those involved. Whether these different democratic cultures, as I want to call them, lead to greater formal democratic participation and active citizenship as Putnam suggests, only further extensive research could reveal. But what is argued here is that the agenda for change that Putnam proposes (and to which I shortly return) may not be appropriate to many of the social movements or new urban cultures of the twenty-first century.

The Question of Difference

Putnam's analysis of social capital pays little attention to questions of difference. If difference is to be taken seriously, understood as embedded in different power relations and contexts, constructed often in an antagonistic relation to others (Mouffe, 1993), then the utopian democratic ideal that underpins Putnam's vision needs to be rethought. As urban populations become increasingly multicultural through processes of globalization, and traditional family households are replaced by complex and changing household forms, and in the context of an ageing population, social relations of race, ethnicity, age and gender need to be centrally addressed in discussions of democratic cultures. Difference is here to stay, and more central to the question of social capital than Putnam is prepared to accept. Thirdly, I want to argue the importance of space in the construction of new democratic cultures, many of which are not necessarily easy to map on to earlier notions of community or civic association. In other words, new forms of social connection are emerging, each with their different spatialities. Finally, taking issue with more static notions of social capital, I want to draw attention to the construction of interests as a fluid and shifting process in the formation of collective identities. These issues are explored in this chapter through three case studies of democratic cultures that have flourished in the late twentieth and early twenty-first century, Putnam's period of social capital decline. These are the Women's Aid/Refuge movement;[1] youth rave cultures;[2] and the University of the Third Age.[3]

    Bowling Alone concludes with an agenda for social capitalists, in an attempt to halt the decline in social connectedness and civic involvement, the so-called 'ebbing of community' (Putnam, 2000: 405). Various policies are proposed: a more family-friendly and community-congenial workplace; improved urban design to facilitate social interaction; new forms of electronic entertainment and communication to reinforce engagement; greater participation in cultural activities; a greater participation in political public life; and a 'new, pluralistic socially responsible great awakening' (Putnam, 2000: 409). Though these may be laudable objectives, my argument is that other new forms of association have emerged in different spaces that may be harder to detect, more informal, and which necessitate different kinds of recognition and politics. In a similar vein, Keith and Pile (1993) highlight how a disengagement with the political mainstream, as demonstrated through the decline in electoral participation, has been accompanied by a rise in alternative forms of political mobilization and involvement around questions of sexuality, race and other identities that challenge conventional notions of the political subject in late modernity. And, as Keith (in Pile and Keith, 1997: 277) points out, these identity politics are highly spatialized.

Before looking at the three cases, let us briefly establish some contours for the analysis of this notion of democratic cultures. As I have suggested, what is missing from the idea of social capital is the idea that all civic engagement, every democratic project, implicitly or explicitly involves some acknowledgement of difference and tension. As Iris Young (1990: 304) has pointed out, the notion of community represents an ideal of shared public life, of mutual recognition and identification that suppresses difference among subjects and groups. In its place she argues for a differentiated citizenship, which attempts to address the problems raised by multiculturalism for a participatory democratic pluralism. Feminists have played a key role in highlighting some of the problems central to pluralism (Phillips, 1991). Small democratic collectives have been criticized on the grounds that they often sustain themselves through processes of homogeneous recruitment, thereby excluding others who differ racially, culturally or sexually. Thus, within feminism the ideal of a unitary and hegemonic form of democratic participation has been dislodged in favour of the idea of a multiplicity of structural and organizational forms and participatory arenas aimed at ensuring inclusiveness and the articulation of a multiplicity of voices (Sirianni, 1993: 307).

Chantal Mouffe (1993) pushes this notion of difference and multiplicity further, arguing that antagonism is crucial to politics. In her formulation, every identity is relational, that is, 'the condition of existence of every identity is affirmation of difference' (1993: 2). The delimitation of a 'we' inevitably involves the creation of a 'them', a process that constitutes all collective identifications. For Mouffe, within this process there is always the possibility that the determination of an 'other' as a 'constitutive outside' will become the site of political antagonism, as the other negates or refuses to recognize our difference. As a consequence, according to Mouffe (1993: 3), the political cannot be restricted to a certain type of institution. Mouffe proposes the concept of a radical and plural democracy, which recognizes the impossibility of a world without antagonism and which does not seek a final resolution of conflicts. In order to radicalize the idea of pluralism, we need to break with rationalism, individualism and universalism, which, she suggests, will enable the multiplicity of forms of subordination that exist in social relations to be exposed and a framework for the articulation of different democratic struggles around gender, race, class, and sex to be constructed (1993: 7).

Difference is embedded in power relations. One place where the idea of social capital falls down is in its lack of attention to the ways in which people involved in different associations, organizations and groups are differently positioned in multiple power relationships which are contingent and shifting and which create new spaces and spatialities and new meanings and identities. For Massey (1999: 284), the question is how 'spatialities of power can be reordered through practices which are more egalitarian, less exploitative and more mutually enabling'. If more fluid notions of power are introduced to analyze the struggle for different identities and political subjectivities, then the notion of resistance can be uncoupled from domination (Keith and Pile, 1993: 2). From such a position emerges the possibility for resistances to be constituted in the spaces and practices of everyday life that may involve symbolic meanings, political discourses, social networks, physical settings, bodily practices and envisioned desires (Routledge, 1999: 69). This perspective allows for different, often hidden or invisible, cultural practices, which may emerge and dissolve in different spaces, places and times, often out of sight, interstitial and fleeting, to be articulated as potential spaces of resistance, democracy and forces for change.

New democratic cultures are constituted spatially and that space makes a difference. Different social groups endow space with a combination of different meanings, uses and values (Routledge, 1999: 70), and spatialities are constituted in the process of performing our various identities (Massey, 1995: 285). The idea of a politics of location is important here, which for Mohanty (1992) describes the historical, geographical, cultural, psychic and imaginative boundaries that provide the ground for political definition and self-definition. For Pile (in Pile and Keith, 1997: 28), the politics of location 'involves not only a sense of where one is in the world – a sense gained from the experiences of history, geography, culture, self and imagination – mapped through the simultaneously spatial and temporal interconnections between people, but also the political definition of the grounds on which struggles are to be fought'. This is a politics that resists the erasure of spaces of difference and differentiation.

Contemporary political interventions, collective identifications and democratic cultures are not static but fluid, unfixed and changing. Here, I want to take issue with Putnam's more stable and fixed picture and argue that community associations and social movements need to be seen as constantly in process, forming, reforming and relocating in new places. The democratic possibilities that may emerge out of these are inevitably unpredictable and contingent. As Laclau (2000: 86) has insisted, 'the only democratic society is one which permanently shows the contingency of its own foundations'. Putnam's account focuses on more formal and fixed forms of social capital, hence its more gloomy conclusions. Elsewhere, I have argued that interests do not exist outside the state pre-formed and defined, but are constructed through interaction with the arenas of the state (Pringle and Watson, 1992, 1996). Similarly, identities are constituted through discursive practices and political struggles. Mouffe (1993: 53) makes a similar point when she argues that different discourses will attempt to dominate the political field and create nodal points through the practice of articulation, but that they can only succeed in temporarily fixing meaning: 'part of the struggle characteristic of modern politics is to constitute a certain order, to fix social relations around nodal points, but successes are necessarily partial and precarious because of the permanence of antagonistic forces'. This is a view that opposes the universalistic realm of modern citizenship that denies difference. It is also opposed to Young's (1990) group differentiated citizenship, where interests and identities are already given.

Finally, Putnam's civic associations are selected to show the links between the falling membership of groups and the decline in participation in formal democracy. Many of the groups, organizations and associations he considers, such as the bowling club, Scouts, Rotary clubs, are very hierarchical and often exclusive. The intrinsic form of these organizations is not therefore of relevance to his argument. It is not the cultural practices of the organization *per se* that leads to greater formal democracy. My interest here, in contrast, is in forms of association which themselves are democratic, what I want to call *democratic cultures*. Just as Putnam argues that the decline in social capital is connected with the decline in formal democracy, I want to suggest that the proliferation of democratic cultures might represent the growth of a different kind of politics of mutual respect, trust and solidarity (all elements of social capital) which may, or may not, translate into formal democracy.

In the next part of the chapter I look at three different democratic cultures, which have thrived in the period of Putnam's diminishing social capital. My hunch is that there are many others. They have been selected among many other invisible and spatially interesting, potentially democratic cultures (for example, the Internet, Alcoholics Anonymous) for three reasons. First, each of these loose associations or groups is in some sense oppositional, and operates according to principles of non-hierarchical participation. Secondly, they represent three often marginalized groups of people (older people, youth, women), each of which is very different culturally from one another. Thirdly, each of these movements has strong spatial elements.

In the analysis that follows, I look at these groups within the framework I have outlined above, which draws attention to issues that are largely absent from Putnam's analysis. The explicit links between these associations and formal democratic participation cannot be drawn without further research. Nevertheless, I do want to suggest that there are many such democratic cultures that have emerged during Putnam's period of social capital decline, and many more will emerge in the future. Given that difference is intrinsic to all social relations and embedded in power, new forms of collective identifications and connectedness will continue to emerge in different spaces, whether visible and invisible, static and mobile. Some of these may enhance a wider involvement in more formal public political life, others will not. It is my contention, however, that associations that are predicated on non-hierarchical and democratic forms of organization are likely to sustain a greater democratic respect and way of being more broadly, even if this does not translate into voting behaviour.

Democratic Cultures

The University of the Third Age

The term 'the Third Age' was first introduced to Britain by Peter Laslett in 1981, when the first British University of the Third Age (U3A) was founded in Cambridge (Laslett, 1996: 3). Laslett argued forcefully that outmoded attitudes, stereotypes and assumptions continued to distort the image of older people, one source of which he saw as the scientific medicine of the early twentieth century. Though not explicitly articulated in the literature, the use of the term 'Third Age' was clearly a discursive strategy to disrupt normative attitudes to older people and aging, and to create a new collective identity. The Second Age is defined as a period where people have full-time employment and family responsibilities, when work is wholly imposed by others and allied with loss of time. This leads, according to Laslett (1996: 188) to the notion that all that people want to do in retirement is take a long rest, the dominant idea in the 1950s being that this was the time to pursue gentle hobbies. With growing longevity, the Third Age for many people may extend for an equivalent period to the Second Age. The Third Age is deployed to replace terms like 'Old Age' to stand for 'dignity and creativity, the social and public significance, the self respect and civic virtue of older people which certainly continue indefinitely into later life' (Laslett, 1995: 10). It is argued that this is a time of personal achievement and fulfilment which is not designated by chronological age.

Another proponent of the Third Age, Eric Midwinter (1992), proposed the idea of the outlawed citizen to describe the exclusion of older people from civic and voluntary life. He identified the many and various forms of such exclusion. For example, until recently the Citizens' Advice Bureau took no one over 65, with obligatory retirement at 70; the Lord Chancellor's department stipulated candidates for magistracy must be below 60 with a preference for under 50; jury service is curtailed at 70. Other enforced retirements included guides and scouts (65); British Red Cross volunteers (75); St Johns Ambulance (no new volunteers over 65); lay magistrates on juvenile panels (65); industrial tribunals (chairperson 72, members 68). Midwinter established that people over 55 were less involved with volunteering than the rest of the population by a ratio of 1:4. Underpinning these exclusions lay the assumption that chronological age can be used as an indication of capacity, denying the obvious diversity in this population. In the place of chronological age, Schuller and Young (1991: 22) argue for new images and metaphors of the life course, suggesting that the medieval use of the circle to depict the life course could replace the image of a line.

The University of the Third Age was initially based on the French L'Université de Tous Les Ages, established in 1972. Founded in 1981, with principles drawn up in the first instance by Laslett, the U3A identified a number of objectives: (1) to educate British society at large about its age constitution and the implications of this; (2) to enhance older peoples' awareness of their intellectual, cultural and aesthetic potentialities and to assail the dogma of intellectual decline with age; (3) to draw on the resources of older people in an affordable way for the development and intensification of their intellectual, cultural and aesthetic lives; (4) to create an institution for these purposes where there is no distinction between those who teach and those who learn, where all labour is freely and mutually offered; (5) to create an institution where learning is pursued with no reference to qualifications and interests are developed for themselves alone; (6) to mobilize university members to broaden the educational opportunities of other older people who want to engage in education; (7) to undertake research on the process of ageing in society; and (8) to encourage the establishment of U3As across the country where conditions permit, and to collaborate with them.

Thus, unlike the Francophone model from which it drew its inspiration, which more closely resembles a major development of university extension work, U3A embodies a culture of mutual aid and voluntarism in adult education where those who teach also learn and those who learn also teach. This is a crucial element of the organization, and evidence suggests that members' confidence, particularly among women, is greatly enhanced through this process. At the same time, U3As make 'older people more visible, creating a public space for generations whose presence can be overlooked and whose worth is often underestimated' (Laslett, 1996: 31). By 1995, 1,500 U3As had been set up in over 20 countries, including 290 in France, 400 in China, 17 in Poland, and 250 in the UK. By 2001 this had expanded to 459 U3As in Britain involving 110,605 people. Local residents set up U3As in local communities, often by word of mouth. Once it is thought that a U3A group has reached its maximum, a new U3A is established. The classes provided depend on local interests and expertise, and can range from architecture, and the history of art, to computing, cooking and poetry reading. Annual fees are kept to a minimum (for example, £20 pounds in Highgate). In Britain, U3As form the intellectual vanguard of the Third Age as a whole. But the organization is also concerned to make a political intervention. As Groombridge put it:

> It has become more and more obvious in our time that government has become impossibly difficult. That is one reason why there have to be Non-Governmental Organizations representing thoughtful, well-informed citizens at major events such as the Rio Summit. The responsibility for policy-making has to be more widely shared. Democracy can no longer be equated with choosing a handful of

people, most of them middle-aged men, most of them no brighter than we are, to decide everything. That is the reason why U3As must get involved in a more deliberate overt way. (Groombridge, n.d.: 29)

There are important spatial elements to the U3A. There is the idea of the learning city, based on the practice that everyone is a learner and teacher and the boundaries between the two are broken down. Many U3As are involved in the exploration of cities, towns and villages as learning units, combining benefits of local knowledge and personal contact with the openness enabled by new technologies (Laslett, 1996: 24). The majority of classes take place in members' homes. They are thus largely invisible except to those involved. This raises issues of class and race-based exclusivity. Members of the U3A are almost exclusively white, often, but by no means entirely, middle-class, with a majority of women. Consider the contrast between a group based in the London borough of Haringey, in Tottenham, and one based in Highgate. The Highgate group is exclusively white and mainly middle-class. The majority of meetings take place in members' houses, although a number of popular courses are run in a local National Trust property, where a small fee is required. The architecture course arranges overseas trips, for which an income of a certain level is clearly a prerequisite. However, the characteristics of this group, when contrasted with that of the Haringey group, highlight the hidden racialization of this otherwise very democratic culture. A politically active Asian woman, who called a meeting in the local Marcus Garvey Library, initiated the Tottenham group in 2000. The initiative was immediately success-ful. Most attendees at this groups' meetings are women, with a majority of these Asian, Afro-Caribbean or African. The Chair of the Committee of the Haringey group, who had come to the UK from Jamaica in the 1950s, com-mented that 'we cannot have the meetings in our houses. Often they are too small, but also for us hospitality is very important. You can't invite people to your home and not give them tea and cakes and other things to eat. The same goes for our Asian members. So if you had the classes at home you'd spend the whole time in the kitchen, or preparing food in advance, and never get to the classes.' The success of this group, then, was dependent on the offer of free, local and accessible municipal facilities where classes could take place. These patterns of inclusion and exclusion appear to have received limited recogni-tion, although this is now changing: 'U3As and similar organisations are springing up throughout the world and part of the strength of that community lies in its diversity. Like all strong communities the differences between the members are not only tolerated but welcomed' (Futerman, 1989: 10).

In summary, the U3A is an association whose collective identity was formed through a number of discursive strategies deployed in mainly local state and community arenas. The members of U3A have constructed an alternative space to the dominant discourses on age. Since its formation, a group of people, particularly women, have come to redefine themselves as 'Third Agers', and have gained confidence, new skills and knowledge as a result. The composition of this association is fluid and growing, its culture deeply democratic and participatory, yet partly through its spatial form it constructs and reinforces exclusions based on race and class.

Dance and Rave Cultures

I will discuss the following two socio-cultural formations, youth dance and rave cultures, and Women's Aid, more briefly, since there has been extensive analysis of both of these movements (see, for example, Garratt, 1998; Reynolds, 1998; Wright, 2001 (on dance culture); and Reinelt, 1995; Rodriguez, 1988 (on Women's Aid)). My reason for discussing them is, first, because they both represent collective social and political intervention that has flourished in the period that Putnam claims has seen a decline in social capital, and, secondly, because their spatial constitution is crucial to the political cultures associated with these practices.

In Britain in the late 1980s and early 1990s, dance music and dance culture (or 'house'), and, in particular, the free party movement, provided an entirely different cultural space, one which had to be fought for, but one which represented a new form of sociality and which was, arguably, also a democratic space. House parties consisting of thousands of people, taking place in different locations, were coordinated covertly with no institutional framework or clear leaders. Gilbert suggests that, after years of Thatcherism and economic decline, 'they provided a source of hope and an outlet for an immense amount of energy released by the breaking down of the social codes of a bankrupt reality' (2001b: 32). Inevitably, the parties came into conflict with the police, resulting in their curtailment, legitimized in terms of a lack of a clear regulatory framework. Dance culture can be theorized as a form of direct action, as opposed to a movement with clear political objectives. The unpopularity of this culture with the forces of law and order was in large part related to its supposedly amoral excess fuelled by the drug Ecstasy.

What specifically interests me here is how the spatial form of 'house' was central to its culture. Gilbert (1996: 26) argues that dance culture 'exploits the power of music to build a future on the desolate terrain of the present. The history of house can be seen as an answer to turn this house into a home.' In the quest to find venues for events, a nomadic structuring of space was thrust on to the early dance scene (Gilbert, 1996: 31). As a result of its criminalization, it became impervious to property relations and erupted wherever a place could be found, breaking down established divisions and zonings of land. Abandoned warehouses were reclaimed and turned to a new purpose. Gilbert describes this as an 'inner-city breakout [which] was not enclosed by boundaries'. The dance-floor is literally carried on the dancer's back, and goes with the dancer wherever s/he goes: it is 'not in space but of space' (Gilbert, 1996; see also Reynolds (1998)). In relation to our earlier schema, space is clearly crucial to this culture's formation, and so too is the notion of fluidity, movement, indeterminacy, and the egalitarianism associated with democratic cultures.

Gilbert (2001b) argues that, before it became commercialized and repackaged as club culture, the 'convergence of early rave's collectivist and anti-authoritarian urges with the anarcho-punk neo-hippies of the free festival circuit produced the most obvious, explicit and self-conscious manifestation of dance culture's utopian ideals: the free party movement'. This was a movement that used informal networking and secret tactics to offer an alternative to elitist and expensive dance venues, in opposition to the state and the population at large. Gilbert (2001b) analyses the strategies of three different groups involved in this culture. The first, *Spiral Tribe*, took their sound system out of the United Kingdom as a result of the restrictions imposed on free parties by the 1994 Criminal Justice Act, travelling to South America, war-torn Bosnia and other places where it was felt there might be a need for the ecstatic anarchy they brought. The second, *Exodus* established a local community at a disused farm squatted near Luton, from which parties and festivals were organized in the locality. And third, Nottingham-based *DIY*, divided their activities between free parties and legal club nights, acting as a bridge between counter-culture and the mainstream. In Gilbert's view, *Exodus's* localism led to an exclusionary politics, which was more concerned with defining the parameters of the community rather than the political task of expanding its frontiers. Nevertheless, it did become a focal point for local community activism. *DIY*'s strategy, he suggests, represented a more political approach in that it defended an alternative cultural space at the same time as extending its scope into mainstream culture. And *Spiral Tribe*, with their free and inclusive parties, succeeded in constituting an alternative public space, rather than just a secret one, though no one could say how many lives were touched by their three-year tour of duty. Gilbert's sympathy, like my own here, is not in establishing whether these cultures are successful in opposing a hegemonic oppressive social order, but rather (and following Laclau and Mouffe, 1985) in thinking about the emergent democratic possibilities opened up by these new antagonisms (see Gilbert, 2001a). And though there has been some concern among feminists that progressive sexual politics find a place in rave (McRobbie, 1994: 168), it has also been argued that rave, with its open displays of happiness, friendliness and auto-erotic pleasure, can 'be read as a challenge to heterosexual masculinity's traditional centrality' (Pini, 1997: 155). What this example emphasizes is the importance of a radical openness to the democratic possibilities of the future, and the notion that these possibilities are unpredictable and never fully constituted.

Women's Aid

Women's Aid was established in the United Kingdom in the middle of the 1970s, with three aims: 'to believe women and children and to make their safety a priority; to support and empower women to take control of their own lives; and to recognize and care for the needs of children affected by violence' (Women's Aid Federation of England, 1998). Since its inception, the number of local refuges in the UK has expanded from three to 250, hardly an indication of a decline in civic engagement, at least by this constituency of women. The refuges are connected through the Women's Aid Federation, which is the key national agency for women and children experiencing physical, sexual or emotional abuse in their homes. Like the other two cultural sites, the space of Women's Aid refuges matters. Visible only to those involved, the refuge is usually an ordinary house in an ordinary street, whose address is kept as secret as possible. This engenders a sense of a secret sisterhood.

The Women's Aid movement is strongly embedded in the ethos of second-wave feminism, which was concerned to redefine the idea of democratic community on more participatory grounds. It emphasized small group support, self-transformation through collective consciousness-raising, the democratization of leadership roles, a dislike of hierarchies, competition and autonomy, and a respect for the experiences of all women. Women's Aid in particular espoused the involvement of residents in the running of the refuge, as a way of providing an alternative to the domination the women have experienced and the passivity of victim status. As time went on, many of the refuges have come to be run by women who had been victims themselves. Participation is seen as empowering and therapeutic, at the same time as providing an alternative to the professional social service model (Rodriguez, 1988). Women from refuges all around the country, workers and residents, come together at regional and national conferences where policies for the national federation are debated and made. For many women, these are key politicizing events from which a broader interest in feminism grew.

Like other arenas of the feminist movement, race and class represented major forces for division and contestation within the Women's Aid movement. Black women challenged white women for speaking on their behalf and for a lack of recognition of the specificity of black women's experience of family, work and welfare. In some instances, this resulted in the formation of separate black and Asian refuges. The Women's Aid movement also confronted many of the problems of participatory democracy first articulated in Freeman's (1975) seminal paper 'The Tyranny of Structurelessness' (first written in 1972). The argument here was that diffuse participatory methods did little to really democratize power. Instead, they often opened up the possibility for elites to emerge informally and for those who wielded the most influence to be even less responsible and less accountable to others. At the same time, the lack of a formal structure made some women feel more inadequate and less empowered as fear of speaking or lack of persuasiveness translated into personal inadequacy (Sirianni, 1993: 290). Thus, egalitarianism bred conformity. When leaders did emerge, they were accused of elitism, yet the inability of the movement to propose women to speak on its behalf gave power to the media to select their own. In the case of Women's Aid in Britain, this was a woman called Erin Pizzey, who had set up the first refuge in Chiswick, but who had no truck with the feminist politics of the Federation. The ideology of structurelessness created the very kind of individualistic star system that it most condemned (Freeman, 1975: 21).

Within this movement, the identity of the battered woman was constructed through intervention into the arenas of the state (see Pringle and Watson, 1992). This identity moved, over more than a decade, from marginal recognition to being inscribed in law and social policies as a mainstream subject of welfare. This was not a pre-constituted subject position with a clear body of interests. Rather it emerged from the shadows through the campaigns and activism of feminists across the world. Over time, authority relations have become more formalized, as policies for state funding, law reform, and educating bureaucrats came to be seen as more effective in empowering greater numbers of battered women. Similarly, as careers opened up for women, feminist activism imagined itself less singularly within an egalitarian form of participatory democracy (Sirianni, 1993: 295). In the United States, the United Kingdom and Australia, for those committed to expanding and stabilizing services, funding came to be seen not as a co-optive trap but as an opportunity to expand services and transform the way agencies defined battering (Watson, 1989). Thus, Sirianni (1993: 297) argues that the 'shelter movement increasingly moved away from democratic collectivist form to create "modified collectives" and "modified hierarchies" that can reap the benefits of more formalized structures while still sharing information'.

Of the three forms of democratic culture discussed here, the Women's Aid movement is now the most incorporated in the arenas of the state. But in many instances, through lack of funding, it remains on the margins of welfare provision, and there is still a lack of adequate services, housing and other forms of provision for battered women. Despite this partial incorporation, Women's Aid remains marked by a politics that sees processes of empowerment and politicization as fundamental. It also retains its spatial distinctiveness as a hidden site, run collectively from within, a place that for residents is temporary and interstitial, where their lives may or may not radically change. This is an indeterminate space for many who pass through, but it is also a democratic space from which new democratic possibilities may emerge.

Conclusion

I have argued in this chapter that though Putnam has clearly identified important shifts in the state of social capital, there may be less cause for pessimism than he suggests. Moreover, the call for a return to formalized participation in civic, professional and other associations may obscure the multifarious communities, associations and forms of connectedness that are already in process. Recognizing these requires looking at different spaces, spaces that may be temporary or barely visible. Once we adopt a more nuanced, spatialized vision of politics, new democratic cultures are revealed which may be just as important as more obvious forms of political association. Just how participation in groups that are democratically organized – in the sense of being non-hierarchical, egalitarian and participatory – connects with a broader sense of mutual respect and solidarity on the one hand, and with formal democratic involvement on the other, remains an open question. It is important, then, to recognize, research and articulate the emergent possibilities of new forms of connectedness and politics that often go unnoticed.

Notes

1 Drawing on my own active involvement over a number of years.
2 Drawing on the work of various cultural analysts (Gilbert, 2001a, 2001b; Reynolds, 1998).
3 Drawing on primary research.

References

Castells, M. (1997) *The Power of Identity*. Oxford: Blackwell.
Freeman, J. (1975) 'The Tyranny of Structurelessness', in *The Politics of Women's Liberation*. New York: McKay.
Futerman, V. (ed.) (1989) *Into the Twenty First Century: A Collection of Lectures and Seminars from the First International U3A Symposium, Cambridge 1988*. Cambridge: University of the Third Age in Cambridge, Symposium Committee.
Garratt, S. (1998) *Adventures in Wonderland: A Decade in Club Culture*. London: Headline.
Gilbert, J. (1996) 'Soundtrack for an Uncivil Society: Rave Culture, the Criminal Justice Act and the Politics of Modernity', *New Formations*, 31: 5–22.
Gilbert, J. (2001a) 'A Certain Ethics of Openness: Radical Democratic Cultural Studies', *Strategies*, 14 (2): 189–208.
Gilbert, J. (2001b) 'Rhizomes and Autonomous Zones: Theorising the Politics of Dance Cultures', in M. Wright (ed.), *Dance Culture: Party Politics and Beyond*. London: Verso.

Groombridge, B. (n.d.) *Emergent Challenges for Universities of the Third Age.*

Keith, M. and Pile, S. (eds) (1993) *Place and the Politics of Identity.* London: Routledge.

Laclau, E. (2000) 'Identity and Hegemony', in J. Butler, E. Laclau and S. Zizek (eds), *Contingency, Hegemony, Universality.* London: Verso.

Laclau, E. and Mouffe, C. (1985) *Hegemony and Socialist Strategy.* London: Verso.

Laslett, P. (1995) 'The Third Age and the Disappearance of Old Age', in E. Heikkinen, J. Kuusinen and I. Ruoppila (eds), *Preparation for Aging.* New York: Plenum Press.

Laslett, P. (1996) *A Fresh Map of Life.* London: Weidenfeld and Nicolson.

Massey, D. (1995) 'Thinking Radical Democracy Spatially', *Environment and Planning D: Society and Space*, 13: 283–288.

Massey, D. (1999) 'Entanglements of Power: Reflections', in J. Sharp, P. Routledge, C. Philo and R. Paddison (eds), *Entanglements of Power.* London: Routledge, pp. 279– 286.

McRobbie, A. (1994) 'Shut up and Dance', in A. McRobbie, *Postmodernism and Popular Culture.* London: Routledge.

Midwinter, E. (1992) *Citizenship: From Ageism to Participation.* The Carnegie Enquiry into the Third Age, Research Paper No. 8. Dunfermline: Carnegie (UK) Trust and Center for Policy on Ageing.

Mohanty, C. (1992) 'Feminist Encounters: Locating the Politics of Experience', in M. Barrett and A. Phillips (eds), *Destabilising Theory.* Cambridge: Polity Press.

Mouffe, C. (1993) *The Return of the Political.* London: Verso.

Phillips, A. (1991) *Engendering Democracy.* Cambridge: Polity Press.

Pile, S. and Keith, M. (eds) (1997) *Geographies of Resistance.* London: Routledge.

Pini, M. (1997) 'Women and the Early British Rave Scene', in A. McRobbie (ed.), *Back to Reality?* Manchester: Manchester University Press, pp. 152–169.

Pringle R. and Watson, S. (1992) 'Women's Interests and the Post-structural State', in M. Barrett and A. Phillips (eds), *Destabilising Theory.* Cambridge: Polity Press.

Pringle, R. and Watson, S. (1996) 'Feminist Theory and the State: Needs, Rights and Interests', in B. Sullivan and G. Whitehouse (eds), *Gender, Politics and Citizenship in 1990's Sydney.* Sydney: University of New South Wales Press.

Putnam, R. (1993) *Making Democracy Work.* Princeton, NJ: Princeton University Press.

Putnam, R. (2000) *Bowling Alone: The Collapse and Revival of American Community.* New York: Simon & Schuster.

Reinelt, C. (1995) 'Moving onto the Terrain of the State: The Battered Women's Movement and the Politics of Contradictory Locations', in M. M. Ferree and P. Y. Martin (eds), *Feminist Organizations.* Philadelphia, PA: Temple University Press.

Reynolds, S. (1998) *Energy Flash: A Journey Through Rave Music and Dance Culture.* London: Picador.

Rodriguez, N. M. (1988) 'Transcending Bureaucracy: Feminist Politics at a Shelter for Battered Women', *Gender and Society*, 2: 214–227.

Routledge, P. A. (1999) 'Spatiality of Resistances: Theory and Practice in Nepal's Revolution of 1990', in J. Sharp, P. Routledge, C. Philo and R. Paddison (eds), *Entanglements of Power.* London: Routledge, pp. 69–86.

Schuller, T. and Young, M. (1991) *Life after Work.* London: HarperCollins.

Sirianni, C. (1993) 'Learning Feminism: Democracy and Diversity in Feminist Organisations', in J. Chapman and I. Shapiro (eds), *Democracy's Place*. New York: New York University Press, pp. 283–312.

Watson, S. (ed.) (1989) *Playing the State: Australian Feminist Interventions*. London: Verso.

Women's Aid Federation of England (1998) Information Leaflet. <http://www.women said.org.uk>

Wright, M. (ed.) (2001) *Dance Culture: Party Politics and Beyond*: London: Verso.

Young, I. M. (1990) *Justice and the Politics of Difference*. Princeton, NJ: Princeton University Press.

12 Spaces of Mobilization: Transnational Social Movements

Byron Miller

We [...] need to re-emphasize reaching out as widely as possible and providing means of participation for as many new people as we can interest. We need to unequivocally understand that our strategic goal isn't to have a small army of courageous, creative, insightful, and bold dissidents. We need many in motion [...] for our goals, thousands and even tens of thousands are still few. To attain needed size and scope, we have to correct the appearance that anti-globalization requires traveling to distant cities and demonstrating in the midst of clubs and tear gas [...] to grow sufficiently to win, our movement needs to offer things for people to do where they live and in accord with their dispositions and possibilities. (Albert, 2001a: 2)

Certainly, local loyalties are central in any emancipatory politics, but solidarity, inter-place bonding and collective resistance demand a decidedly scaled politics that can challenge the totalizing powers of money and commodified culture and provide a credible alternative. What is disturbing in contemporary politics of resistance is not that the paramount importance of scale is not recognized, but rather that oppositional groups have failed to transcend these confines of a 'militant particularism' or 'particular localism'. (Swyngedouw, 1997a: 175)

Social movements are essential to any well-functioning democracy. While electoral and judicial processes render key governance and regulatory decisions at specific moments, democracy is fundamentally an ongoing communicative process. Political parties give voice to a number of groups regarding a limited range of issues and positions, but much remains outside mainstream political discourse – issues and positions too controversial or threatening to powerful interests to be championed by mainstream political parties. Social movements give voice to people and causes outside the established power structure, and through ongoing discussion, education, and mobilization, create the conditions and pressures necessary for broader debate and action within the official institutions of democracy.

The logics and processes of social movement mobilization are multiple. Appeals to a variety of forms of collective identity, self-interest, and altruism all come into play when social movements mobilize to resist oppression and exploitation. While social movement mobilization has always been a complex amalgamation of processes, the level of complexity as well as the strategic dilemmas that activists face have reached new extremes as social, economic, and political relations become more globally integrated.

Globalization implies, first and foremost, a change in spatial organization and order, and in the associated 'geometries of power' (Massey, 1994). Globalizing processes can clearly be seen in the economy, as transnational processes of investment, production and trade become functionally integrated on a global scale (Dicken et al., 1997). A parallel process operates in the political realm as nation-states are 'hollowed out', their capacity for regulating increasingly mobile capital reassigned to global regulatory institutions such as the World Trade Organization, the International Monetary Fund, and World Bank, and to local institutions (Jessop, 1994). In the cultural realm, globalization occurs through the transnational proliferation of commodity relations, media images, and entertainment (Featherstone, 1990), as well as through transnational migration, communication, and cosmopolitan elite interactions (Appadurai, 1993; Hannerz, 1996).

Globalization has significant implications for social movement mobilization. Traditional geographies of mobilization, rooted in localized places and the nation-state, appear to be in relative decline. Castells (1996: 412), for instance, argues that the 'space of places' is being replaced as the 'space of flows', with the logic and meaning of places being 'absorbed in the network'. Instead of mobilization rooted in place and territory, new geographies of mobilization seem to be in ascendancy. The rise of global civil society (Boli and Thomas, 1999) and the activities of tens of thousands of transnational NGOs and anti-globalization social movement organizations (Smith et al., 1997; Keck and Sikkink, 1998; Mittelman, 2000; Edwards and Gaventa, 2001) suggest a fundamental shift away from place-based forms of political organizing and towards transnational mobilization networks. Indeed, the recent massive anti-globalization protests in Seattle, Prague, Washington DC, Genoa, and Quebec City have demonstrated the ability of activists to communicate, organize, and travel on a global scale. Capital no longer unambiguously commands space to the relative exclusion of workers.

Yet despite the success of anti-globalization protests in calling attention to the deprivations of global neo-liberalism, very little has actually changed in either the behaviour of transnational corporations or the institutional structures of global governance. From the perspective of progressive social movements, the work of building an emancipatory world order has just begun. Objectives and strategy are being hotly debated (Albert, 2001a, 2001b) with many asking 'why things aren't accelerating as we might have hoped?' (Albert, 2001a: 2). A major part of that debate centres on spatial strategy and tactics. A central question is whether to focus on global organizing and protest (e.g. demonstrating at the global summits of world leaders and global regulatory institutions), or to try to build a deeper base of support through local organizing. Social movement strategists engage the global/local dilemma on an ongoing basis, but the distinction is rarely straightforward. As the work of Smith (1992, 1993), Massey (1994), Agnew (1996), Harvey (1996, 2000), and Swyngedouw (1997a, 1997b) demonstrates, localized places and global spatial processes are not the bounded, either/or alternatives that Castells' formulation implies. Rather, the local and the global are deeply intertwined' (Swyngedouw, 1997b: 137) and exist in dialectical relation to each other. As Massey observes:

> The geography of social relations is changing. In many cases such relations are increasingly stretched out over space. Economic, political and cultural social relations, each full of power and with internal structures of domination and subordination, [have] stretched out over the planet at every different level, from the household to the local area to the international [...]. It is from this perspective that it is possible to envisage an alternative interpretation of place [...], constructed out of a particular constellation of social relations, meeting and weaving together at a particular locus. (Massey, 1994: 154)

From this perspective, the strategic issue becomes less a question of whether mobilization activities should have a global or local focus, and more one of where best to break into 'glocal' relationships. After all, transnational corporations as well as the activists who oppose them remain, for the most part, embedded in place-based relationships even as they coordinate activities across space. Just as transnational corporations have learned to base their global accumulation strategies on the identification of favourable locations for production, reproduction, and marketing, so too, social movements are learning to identify, negotiate with, and build alliances among people located in diverse places that are nevertheless bound up in complex socio-spatial processes operating on a global scale.

Collective Identity and the Spaces of Mobilization

All social movements face the same fundamental problem: how to mobilize widespread support and participation. At the root of the problem lies what Olson (1965) identified as the free-rider dilemma: why would a person participate in collective action when s/he could reap the benefits of others' participation without actually becoming involved? Olson's solution to the free-rider problem – that people participate when offered selective incentives (e.g. money, coffee and donuts) – has been shown to be highly implausible (Miller, 1992, 2000). Yet the question Olson poses is crucial. Most activists would concur that the biggest task any movement faces is the mobilization of support and participation. Most people who share affinities with social movements never participate in them.

While there is no single answer to the problem of free-riding, a very important part of the answer lies in collective identity. Numerous social movements theorists (e.g. Jencks, 1979; Calhoun, 1988; Mansbridge, 1990; Melucci, 1994) have argued that collective identities are just as fundamental to a sense of selfhood as individual identities. The fact that we identify with others allows us to 'redefine our "selfish" interest so that it includes our subjective understanding of the interests of a larger collectivity of which we are a part' (Jencks, 1979: 54). In short, our collective identities lead us to think in terms of the collective good, so that the interests of others become our own. Approached from this perspective, participation in social movements should come as no surprise. Of course, other factors also come into play. Collective interest may conflict with individual self-interest; the strength of collective identities and bonds may vary substantially among individuals, groups, and places; political opportunities may be so limited, or the risks of political activism so high, that any venture into activism may seem futile (Tilly, 1978; Tarrow, 1998; Miller, 2000). The internal balance of individuals' multiple identities and interests, the strength of relations with others, and place-specific threats and opportunities factor heavily into decisions to participate in collective action.

For any given issue, social movement mobilization requires appeals to particular social collectivities, building alliances among collectivities, and selection of specific targets that compellingly represent a source of oppression and/or render it vulnerable. While David Smith (2000: 1151) is undoubtedly correct when he observes that 'some of the greatest struggles for social justice in recent history (for example, for black civil rights in the USA and against apartheid in South Africa) were more a case of the universalist notion of equal moral worth countering particular social constructions of difference', it is crucial to recognize that the goals which movements strive for are something quite different from the social collectivities they mobilize. For instance, in his seminal analysis of the US civil rights movement, McAdam (1999) identifies middle-class black churches, black colleges, the Southern Christian Leadership Conference, and the National Association for the Advancement of Colored People as the movement's key mobilizing structures. Almost all social movements appeal to specific collectivities (even if their objectives are to improve the broader human condition), usually through key social organizations representing those collectivities:

> It is a challenger's capacity to appropriate sufficient organization and numbers to provide a social/organizational base – and not that organization itself – which makes mobilization possible. Would-be activists [...] must either create an organizational vehicle or utilize an existing one and transform it into an instrument of contention. (McAdam et al., 2001: 47)

McAdam, Tarrow and Tilly explain that in the course of mobilizing local networks of black churches, activists 'had to engage in creative cultural/organizational work, by which the aims of the church and its animating collective identity were redefined to accord with the goals of the emerging struggle' (2001: 47). Examples from civil rights and other movements demonstrate that while identities may never be fixed, appeals to already-existing identities are nonetheless crucial. All movements interpret and assign meaning to social practices and conditions in attempts to 'mobilize potential adherents and constituents, to garner bystander support, and to demobilize antagonists' (Snow and Benford 1988: 198). The ways in which a movement frames its messages must resonate with the movement's target audiences, and this is largely a function of the empirical credibility of messages (their fit with events), experiential commensurability (experiences of the target audiences), and narrative fidelity (resonance with cultural narratives and myths) (ibid.: 208). Such characteristics can vary tremendously over space. It should come as no surprise, then, that social movement mobilization exhibits substantial spatial variation.

How, then, do movements ever mobilize beyond place-specific social realms? Much of the answer lies in the *framing* of movement messages. Harvey (1996), drawing on the work of Raymond Williams, argues that 'militant particularism' plays a crucial role. Militant particularism begins with place-specific experience, but goes far beyond the bounds of place as 'ideals forged out of the affirmative experience of solidarities in one place get generalized and universalized' (Harvey, 1996: 32). According to Harvey, 'Williams appears to suggest that many if not *all* forms of political engagement have their grounding in some kind of militant particularism' (ibid.). But extending solidarity and engagement beyond the local realm can prove to be problematic:

> The move from tangible solidarities understood as patterns of social life organized in affective and knowable communities to a more abstract set of conceptions that would have universal purchase involves a move from one level of abstraction – attached to place – to another level of abstraction capable of reaching out across space. And in that move, something [is] bound to be lost. (ibid.: 33)

Reaching out across space necessarily involves the loss of place-based solidarity and the introduction of a differentiated politics. This is reflected in the framing of movement messages:

> Even the language changes, shifting from words like 'our community,' and 'our people' in the coalfields to 'the organized working class,' the 'proletariat' and the 'masses' in the metropolis where the abstractions are most hotly debated [...]. The shift from one conceptual world, from one level of abstraction to another, can threaten the common purpose and values that ground the militant particularism achieved in particular places. (ibid.)

Harvey and Williams call attention to the declining resonance of movement messages as they become further removed from the identities and everyday concerns of place-specific collectivities. This represents the central challenge that transnational social movements face. Transnational social movements must find ways to frame and reframe broad messages so they will resonate with a diverse and fluid array of collectivities in a wide range of place-specific and not so place-specific circumstances.

Fluid Collective Identity Formation

Cultural homogenization is perhaps the best known of the cultural globalization hypotheses, popularized through Barber's book *Jihad vs. McWorld* (1995). Barber argues that one very real scenario facing the world involves 'onrushing economic, technological, and ecological forces that demand integration and uniformity and that mesmerize peoples everywhere with fast music, fast computers, and fast food – MTV, Macintosh, and McDonald's – pressing nations into one homogenous global theme park, one McWorld tied together by communications, information, entertainment, and commerce'. Barber's McWorld nightmare is based both in time–space compression (Harvey, 1989) and the colonization of the lifeworld (Habermas, 1984, 1987). In this world, diverse cultures are not only bombarded and overwhelmed by Western cultural images, but much of culture – defined as collective projects to give the world meaning – is itself supplanted by commodity relations. The likely response, Barber fears, is

> the grim prospect of a retribalization of large swaths of humankind by war and bloodshed: a threatened balkanization of nation-states in which culture is pitted against culture, people against people, tribe against tribe, a Jihad in the name of a hundred narrowly conceived faiths against every kind of interdependence, every kind of artificial social cooperation and mutuality: against technology, against pop culture, and against integrated markets; against modernity itself as well as the future in which modernity issues. (Barber, 1995: 1)

In both 'Jihad' and 'McWorld', Barber sees a common lack of democratic deliberation, consensus and guidance. Neither reactionary anti-modernist movements nor ever-deepening commodity relations have a basis in broad democratic decision-making.

Barber's analysis overlooks several key aspects of the cultural dynamics of a globalizing world. Reactions to the commodification and rationalization processes of modernity include not only reactionary, defensive movements but also progressive, emancipatory ones (Habermas, 1987). Indeed, one aspect of rationalization is that more and more aspects of everyday life are opened up to deliberation and debate: it can be argued that the potential for bringing more aspects of cultural and economic life under democratic control is actually expanding. Yet the most significant problem in Barber's analysis concerns the homogenization thesis itself. While there is little doubt that cultural homogenization is one dimension of globalization, it can be argued that 'the central problem of today's global interactions is the *tension between cultural homogenization and cultural heterogenization'*. The global proliferation of Western commodities and culture is not a simple one-way flow. Commodities and culture are adopted and adapted in place- and culturally-specific ways, becoming 'indigenized' in ways that alter the cultural meaning and significance associated with their original context. Moreover, cultural flows occur in complex overlapping networks that significantly deviate from simple centre– periphery relationships. In a globalizing world, homogeneity and heterogeneity exist in a non-zero sum, dialectical relationship:

> The globalization of culture is not the same as its homogenization, but globalization involves the use of a variety of instruments of homogenization (armaments, advertising techniques, language hegemonies, and clothing styles) which are absorbed into local political and cultural economies, only to be repatriated as heterogeneous dialogues of national sovereignty, free enterprise and fundamentalism. (Appadurai, 1993: 287)

If Appadurai is correct, it would seem there is little danger of a global loss of cultural diversity, as some homogenization arguments imply. But this does not mean that existing cultural communities and traditions are not threatened. On the contrary, the boundaries of cultural communities have become increasingly porous to global flows of all kinds, undermining their stability. While it may previously have been possible for individuals to 'live in ignorance of most of the combined cultural inventory of the world, in the global ecumene each one of us now, somehow, has access to more of it – or, conversely, more of it has access to us, making claims on our senses and minds' (Hannerz, 1996: 25). How we respond to these claims can vary dramatically.

Drawing on the work of Robert Merton, Hannerz (1996), King (1997), Cheah and Robbins (1998) and others have drawn a direct connection between globalization and the rise of cosmopolitanism. Genuine cosmopolitanism is characterized by openness to Others and a desire to participate in other cultures. It requires a wide-ranging cultural competence and flexibility and is generally associated with transnational professional and intellectual occupations. The growth of transnational cosmopolitan networks and an associated *xenophilic* outlook is crucial to the growth of transnational civil society and transnational social movements (Keck and Sikkink, 1998; Boli and Thomas, 1999). But while a growing proportion of humanity is regularly engaged with multiple cultures, this does not necessarily produce cosmopolitanism. Much of the world's transnational interaction stems not from an openness towards and appreciation of Others, but from the logic of globalizing trade and investment. For many engaged in transnational interaction, 'travel is ideally home plus more and better business' (Hannerz, 1996: 104). Even further from the cosmopolitan outlook are the xenophobia and 'heightened forms of separatism and distancing with reference to relations with the Other [that] are tending to occupy a central place in the politics of identity in various regions of the world' (Marden, 1997: 58). These anti-cosmopolitan forms of identity politics can take a variety of forms, from nationalism and other forms of provincialism, to racism, ethnic discrimination, sexism, homophobia, and religious fundamentalism. This variety of res-ponses suggests a 'precarious balance between "global flows" and "cultural closure". There is much empirical evidence that people's awareness of being involved in open-ended global flows seems to trigger a search for fixed orientation points and action frames, as well as determined efforts to affirm old and construct new boundaries' (Meyer and Geschiere, 1999: 2). Meanings and identities are socially constructed; they are not immune to the varying patterns of flow and fixity in the economy or the actions of the state designed to engender place-specific identity and loyalty. A central problem for social movements in a globalizing world becomes how to 'grasp the flux' of meanings and identities (Hannerz, 1992).

Meanings and identities have never been fixed and static. Nor are they completely free-flowing under contemporary conditions of globalization. The global flow of commodities, messages, and meanings creates an incredibly wide range of cultural and life-choice possibilities in a greatly expanded realm of the imaginable. Yet, with the exception of cosmopolitans who may form relationships in and travel among diverse places around the world, the daily time–space paths of most people are distinctly local. The local remains the realm in which global 'flows' of all kinds are received, interpreted, and adapted. As Hannerz (1996: 28) argues, the principle vehicles of cultural production and distribution 'remain those which we largely take for granted as parts of local life [...]. The everyday and face-to-face may be small-scale; in the aggregate, it is massive'. Or, simply put, 'culture sits in places' (Escobar, 2001). Those places may have contested, blurred and shifting boundaries, and may exist at a variety of scales, from the nation-state (which may expend considerable effort towards fixing national identity) to the home (where identities and meanings are contested on a micro-scale), but they retain boundaries nonetheless.

Clearly, cultural globalization does not produce homogenization, but rather 'an organization of diversity, an increasing interconnectedness of varied local cultures, as well as a development of [cosmopolitan] cultures without a clear anchorage in any one territory. And to this interconnected diversity people can relate in different ways' (Hannerz, 1996: 102). Social movements must relate to this diversity in a multitude of ways if they are to have any prospect of building broad alliances. Shared culture and collective identity are the basis of most forms of social movement mobilization, yet the diversity and flux of a globalizing world makes finding such commonalities, and holding alliances together, increasingly problematic.

A variety of complex processes produce a maze of cultural disjunctures. Following Appadurai (1993), cultural disjunctures increasingly arise around five axes:

1 Ethnic differentiation associated with highly mobile populations including 'tourists, immigrants, refugees, exiles, guestworkers and other moving groups and persons [who] affect the politics of (and between) nations to a hitherto unprecedented degree' (1993: 276);
2 Fluid global configurations of technology that 'now [move] at high speeds across various kinds of previously impervious boundaries' (ibid.: 277);
3 Rapidly shifting patterns of capital investment and disinvestment;
4 'The distribution of the electronic capabilities to produce and disseminate information [providing] large and complex repertoires of images, narratives and [ethnic landscapes] to viewers throughout the world, in which the world of commodities and the world of news and politics are profoundly mixed [blurring] the lines between the realistic and the fictional' (ibid.: 278);
5 'Concatenations of images [that] are often directly political and frequently have to do with the ideologies of states and the counter-ideologies of movements explicitly oriented to capturing state power or a piece of it' (ibid.: 279).

The disjunctures of an increasingly fluid cultural realm make sustained mobilization and alliance-building that much more difficult. The cultural flows identified by Appadurai result in a variety of semantic and pragmatic problems. Meaning is rarely clear outside context; understanding the meanings of the images and messages propagated through these flows requires translation into a multitude of new contexts. Even if meaning can be reliably translated, the actions that flow from them is dependent upon place-specific conventions. Given these complexities, it is difficult to see how social movements can organize beyond place-specific contexts. Yet transnational social movements must and do.

Mobilization Dilemmas in a World of Cultural Flow

Under cultural globalization, the stability of many forms of collective identity becomes precarious. Global flows of messages and images, transnational migrations, and the creolization of culture reinforces and accentuates the fluidity of identity formation. In a world of global cultural flows greater scope for choice enters into individual adoption of collective identity. At the extreme, we face the 'postmodern nightmare' of unlimited flux and diversity, with nothing more than fleeting, ephemeral commonalities (Kobayashi, 1997). Under such an extreme we would undoubtedly witness the near-death of all forms of collective political action. But this is clearly not the case. The destabilization of boundaries – both social and geographical – should not be equated with their disappearance: 'While slogans such as "There are no borders" characterize the rhetorics of multinationals who seek to advertise their products all over the world, global flows actually appear to entice the construction of new boundaries as much as the reaffirmation of old ones' (Meyer and Geschiere, 1999: 5). Efforts to 'fix' the 'flow' of social and spatial boundaries of collective identity are ongoing, and they occur from both above and below. The primary actor working to fix boundaries from above is undoubtedly the nation-state. The nation-state's boundaries have become increasingly permeable to transnational capital flows that render economic processes less predictable and even anarchic, in turn destabilizing communities and regions that depend upon stable patterns of investment. But the nation-state's ceding of sovereignty in the economic realm does not necessarily validate the 'weak state' thesis (Weiss, 1998). Marden, for instance, argues that the state has been 'reconstituted', not weakened, and that it has the ability to

> utilize new techniques and technologies of surveillance and control; indeed the state, with its abstraction of power across time and space, now has the capacity, in Habermasian terms, to colonize the social world more intensely than it ever had in the past [...]. The state, therefore, is not withdrawing from the lifeworld, nor is there a decentralization and redistribution of power downwards to subnational localities and communities. (Marden, 1997: 58)

While the nation-state may have the greatest capacity to construct a common collective identity, a variety of other efforts – which usually overlap rather than challenge national identity – are undertaken from below. Many forms of identity construction from below are, not surprisingly, place-bound. In a neo-liberal world in which national policies do little to ensure distributive and territorial justice, place-based communities increasingly find themselves in competition with each other. A variety of local and regional forms of identification tend to arise in response to place-specific forms of economic deprivation (Markusen, 1987; Cox and Mair, 1988; Hudson, 2001). Nationalism may also be harnessed from below, sometimes to progressive ends such as resistance to IMF-imposed austerity measures, but also for reactionary ends, such as the anti-immigrant scapegoating campaigns of Jean-Marie Le Pen and Patrick Buchanan.

The political dilemmas posed by fragmented, multiple, and diverse collective identities are well known. Differentiated collective identities give rise to a diverse array of political movements and demands, while the structures of the state and capitalism exhibit considerably less fragmentation. Harvey (2000: 71) summarizes the problem: 'The moves of one oppositional or protest movement can confound and sometimes check those of another, making it far too easy for capitalist class interests to divide and rule their opposition.' One response has been to look for a broad global identity that can serve to unite diverse cultural and place-specific collectivities. Harvey, along with others like Gitlin (1995) and Hobsbawm (1996), has argued that class is the common master-narrative that can unite the diverse interests of oppressed collectivities. But while every individual does indeed occupy a class position, and most collective identities can be associated with a limited array of economic roles, it does not necessarily follow that oppressed collectivities will view their oppression in class terms. On the contrary, many collectivities – ethnic, gender, regional, sexual – view their oppression primarily in terms of a lack of recognition and an attendant denial of full citizenship rights Honneth and Anderson, 1996. Moreover, working-class communities across the globe not only struggle against capitalists, they also compete against each other for employment and investment. So while uncovering 'the class content of a wide array of anti-capitalist concerns' (Harvey, 2000: 81) may be an effective mobilization strategy for many issues in many places, it also has significant limitations.

A parallel search for a broad, unifying collective identity has occurred in the cultural realm. With the acceleration of cultural globalization and the emergence of transnational cosmopolitanism, analysts such as Meyer et al. (1997) and Boli and Thomas (1999) have argued that a world culture has emerged in what they term the 'world polity'. They assert that similar conceptions of states and citizenship rights have become common across the world and attribute these commonalities to the transnational diffusion of Western values. Boli and Thomas's underlying thesis is that humans everywhere have similar needs and desires. They are capable of acting in accordance with common principles of authority and action, and share common goals. In short, human nature, agency, and purpose are universal, and this universality underlies the many variations in actual social forms (1999: 35).

But to the extent that there is a 'world culture' in the 'world polity', it is certainly one of cosmopolitan elites operating in the transnational circuits of civil society, state, and corporate structures and not generalizable to diverse populations across the globe. As Hall (1997: 67) observes, 'what we call "the global" is always composed of varieties of articulated particularities [...]. The global is the self-presentation of the dominant particular.' The cosmopolitan/world culture thesis, it would seem, suggests only a very limited potential for the construction of a common world collective identity and culture around which movements might mobilize.

Much of the debate over 'master frames' for broad-based mobilization can be traced back to debate over the nature of a just world order. Positions in the debate usually fall under either the 'politics of cultural recognition' or the 'politics of material distribution'. Nancy Fraser's treatment of the topic is perhaps the most comprehensive and synthetic, given her even-handed attention to both the material and cultural dimensions of justice. Unlike many proponents in these debates, however, Fraser considers the material/culture dualism to be false. Instead, she asserts that 'justice today requires *both* redistribution and recognition' (Fraser, 1995: 69). The axes of injustice that Fraser points to, parallel two basic forms of globalization – economic and cultural – and help us to think about the ways in which globalization processes can be unjust. Globalization can produce various forms of socio-economic injustice, including exploitation, economic marginalization, and deprivation. It can also produce cultural injustice, including cultural domination, non-recognition, and disrespect. While these forms of injustice can be separated analytically, in practice 'the two are intertwined. Even the most material economic institutions have a constitutive, irreducible cultural dimension; they are shot through with significations and norms. Conversely, even the most discursive cultural practices have a constitutive, irreducible political-economic dimension; they are underpinned by material supports' (ibid.: 72). Fraser's formulation identifies the forms of grievance that may arise from globalization processes, and, in turn, conceptions of justice around which transnational social movements might organize. Her formulation also clearly points to the intertwining of the cultural and the political-economic, suggesting that multiple issue frames – economic justice and cultural justice – may be effective bases for mobilization. Avoiding the material/cultural, either/or strategies of those seeking a master mobilization frame leaves conceptual room for addressing collectivity-specific concerns. Yet despite the fact that her analysis illuminates multiple forms of grievance as well as remedies that may produce a more just world, its abstract character leaves little room for recognition of the diverse geographies of grievance. We are still left with the dilemma of how to build alliances among geographically and socially diverse collectivities.

If there is no master frame for a broad unifying collective identity in either the economic or cultural realms, how might emancipatory movements appeal for broad support? David Smith (2000) eschews notions of universal cultural values, instead stressing the 'recognition of human sameness'. By this he means that all human beings have common physiological, psychological, and social needs, which are given specificity in particular social and cultural contexts. For Smith, extending the spatial scope of beneficence

> will come about neither [by] external forces nor something internal to the self, neither [by] the abstract reason of intellectual persuasion nor the immediate experience of empathy with close people [...] it must have something to do with understanding what suffering actually means, arising from a combination of personal experience and the imaginative capacity to generalize from the experience of others. (Smith, 2000: 1158)

Smith argues that we extend our concern to others not based on a common cultural/collective identity, but rather based on our common humanity. This is solidarity in the truest sense of the word – based not on an identity of which we are part, but on the desire to alleviate the suffering of others with whom we share no collective identity, other than to be human. But Smith also recognizes that 'the experiential basis of empathy [...] is likely to be local' (2000: 1159) and that 'transcending local partiality' is part of a maturation process that cannot be taken for granted. Feeling moral responsibility for and, in turn, taking action in solidarity with distant Others requires 'an understanding of the way[s] in which [their lives] are bound up with our own, in the globalizing world economy' (ibid.). Such understanding is often difficult to achieve.

There are clearly multiple bases for collective political action. Appealing to common class identity, common cultural identity, or a sense of common humanity may all be effective means for social movements to mobilize support and participation. Yet each has significant limitations. The fact remains that for most of humanity, the bulk of everyday life experience occurs in place-specific contexts where most strong bonds, as well as understandings of interrelationships, are developed. Except to a growing cosmopolitan class whose members interact in transnational networks, appeals to 'world culture' values and identities are likely to seem far removed from everyday reality. The central challenge for most social movements, then, is not to find a grand unifying basis for mobilization, but to appeal to multiple scale- and place-differentiated collectivities. Or, more specifically, to find flexible and nested ways to construct and frame issues so they resonate among different groups in different places.

Building Transnational Alliances in Place and Across Space

While most movements would undoubtedly prefer to mobilize around one clearly defined set of concerns, norms, and identities, the potential for master frames that will resonate across the globe is quite limited. As Khagram, Riker and Sikkink (2002: 13) point out, there are very 'few examples [...] of truly transnational collective identities' with which abstract universal messages might resonate. Because collective identities are constructed in place and space, they inevitably take on place- and space-specific form, with associated concerns and norms. As Tilly (2003: 222) observes:

> collective identities activated in contentious politics vary along a continuum whose poles we can call embedded and detached. At the *embedded* pole we observe clumps of relations, representations, and understandings that pervade a wide range of routine social interaction as well as forming the bases of collective claim making. Under most circumstances the identities woman, Nahuatl-speaker, neighbour, and peasant, fall toward the embedded end of the continuum. At the *detached* pole we observe clumps of relations, representations, and understandings that constitute identities in contentious claim making but rarely appear explicitly in routine social interaction. Under most circumstances the identities citizen, worker, Mexican, and socialist, fall toward the detached end of the continuum.

While one may offer alternative examples of embedded and detached identities (e.g. 'woman' may also operate as a detached collective identity in the transnational women's movement; or 'socialist' may be a highly embedded identity in the politics of Raymond Williams' Welsh mining towns), the principle Tilly lays out is surely correct: different bases of spatial interaction produce different types of collective identity which in turn present different mobilization challenges. Elaborating on this point, Tilly examines the ways in which differing relationships between spatial mobility and spatial proximity produce varying degrees of 'modularization and detachment'. I want to take his general framework in a slightly different direction, examining the ways in which relationships between

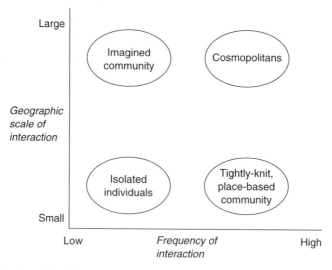

Figure 12.1 Spatial relations and collective identity formation

scale of interaction and frequency of interaction relate to different types of collective identity (Figure 12.1). Much of the contemporary identity literature addresses the growing prevalence of cosmopolitan identities: these tend to be formed in the context of frequent, large-scale, transnational patterns of social interaction that often serve to expand tolerance, cultivate cultural competence, and develop an appreciation for other cultures. At the opposite extreme is a pattern of relatively infrequent social interaction that, when it occurs, is extremely localized; this pattern tends to produce a very parochial and individualistic form of identity. Much of the literature on collective political action stresses the importance of tightly-knit, place-based communities: these are characterized by frequent but highly localized social interactions that breed familiarity and strong, but highly localized, bonds. And finally, virtually everyone in modern societies is connected to others beyond the realm of everyday social interaction, if only through electronic and print media. Actual face-to-face interaction with those distant others, however, is typically infrequent. The paucity of interactive communication means that knowledge and direct bonds are weak, but the stories and myths of commonality may be strongly promoted by mass media and the state. This large-scale, low-interaction collective identity is a form of imagined community (Anderson, 1991).

Of course, these forms of collective identity are not mutually exclusive. Cosmopolitans operate in transnational circuits and may develop cross-cultural, hybrid identities, but most still have a specific place to call home and may be well integrated into a place-based community. Many members of tightly-knit, place-based communities may also accept and deeply identify with the narratives and mythology of a broader-scale collectivity of which they have little first-hand experience. Even isolated individuals who rarely engage in meaningful interaction with other human beings may strongly identify with an imagined community; such a community may actually fulfil, or at least substitute for, a deeply-felt need for communion with others not satisfied in the context of daily interaction. Collective identity, then, is typically complex and multiple, organized along lines that simultaneously involve social and spatial cleavages.

The norms of a collectivity specify appropriate everyday behaviour as well as its highest values and objectives. Any social movement must frame its message to resonate with collectivities' norms, and this almost always entails a degree of cultural specificity. Among cosmopolitans participating in a growing 'world culture', for instance, abstract universal concepts such as 'human rights' are held up as central values. Human rights are broadly defined in documents such as the United Nations Universal Declaration of Human Rights, but when applied to specific societies disputes inevitably arise over civil versus economic rights, the rights of individuals versus cultural preservation, and so on. Addressing such disputes, the Executive Board of the American Anthropological Association concluded in 1947 that: 'Only when a statement of the right of men [*sic*] to live in terms of their own traditions is incorporated into the proposed Declaration, then, can the next step of defining the rights and duties of human groups as regards each other be set upon [a] firm foundation' (cited in Harvey, 2000: 88). In other words, in practice, human rights are culturally specific.

If a broad concern with human rights underlies the normative framework of cosmopolitans, then diametrically opposed would be an emphasis on individual rights to the exclusion of any broader collective concerns. Such an extreme position would be associated with isolated individuals lacking meaningful social bonds. Distinct from these polar opposites is a multitude of collective identities centred on place-based and imagined communities. Their cultural expressions of norms and values are potentially as numerous as the collectivities themselves. This diversity underscores the daunting challenge faced by transnational social movements. They must attempt to harmonize broad universal objectives with a multitude of culture- and place-specific norms, values and issues.

One very important dimension of this universal–particular dialectic centres on struggles over the production of scale (Smith, 1992, 1993; Delaney and Leitner, 1997; Swyngedouw, 1997a, 1997b). Struggles are never inherently local, regional, national, or global in scale. Scales of struggle are, rather, socially and politically constructed. Shifting the scale of struggle (often through struggle) can result in a gain or loss of legitimacy, resources, or resonance. As Harvey points out, one of the key challenges facing all but the most localized movements is 'how to connect political activities across a variety of geographical scales' (2000: 83).

The framing of messages to resonate at the strategically optimal scale(s) of struggle is imperative to any successful mobilization effort. For Harvey, the Zapatistas are perhaps the ultimate scale-savvy movement. Their ability to 'transform what is in effect a local struggle with particular issues [...] onto a completely different scale of analytics and politics [is what] has made the uprising so visible and so politically interesting' (2000: 80). Harvey observes that their claims

> rest firmly on local experience but operate more dialectically in relation to glob-
> alization. They appeal, for example, to the embeddedness of local cultural forms
> but also use the contradiction implicit in the current acceptance throughout the
> world of certain norms [...]. Globalization implies widespread acceptance of cer-
> tain bourgeois notions of law, of rights, of freedoms, and even of moral claims
> about goodness and virtue. (Harvey, 2000: 85)

But straightforward appeals to law, rights, or freedoms in the abstract are unlikely to resonate beyond the cosmopolitan elite. Such broad frames do little to connect directly to the collective identities and norms that guide most people's lives. To mobilize broad international support localized movements must engage in a scale-sensitive form of frame-bridging that can 'amplify, interpret, and legitimate local claims by appealing to international norms. [Such movements] can then use the international arena as a stage or mirror to hold state and international organization behavior up to a global judgment about appropriateness' (Khagram et al., 2002: 16).

The Zapatistas have indeed been brilliant at this sort of scalar reframing. As Routledge (1998) points out, there have been numerous peasant uprisings in Chiapas, but the Zapatistas were the first to mobilize international support. The Zapatistas reformulated traditional peasant demands for land reform and control of the means of production to include broad ecological themes, calls for women's rights and gay rights and, perhaps most importantly, opposition to NAFTA just as it was being implemented. As Froehling observes, the Zapatistas' demands 'were not of course "just symbols", but very real issues that could be taken up and related to the situation at home by people everywhere. The emphasis on women's rights, in conjunction with statements about gay rights and other prominent issues managed to convey the image of a guerrilla [movement] that was fighting for every just cause possible' (1999: 174). Moreover, the romantic image of the highly literate, rebel mestizo spokesperson, Subcommandante Marcos, provided 'many of the signifiers that [helped to] knit together a wide coalition of supporters outside Chiapas [and to make] connections to other social movements that have very little to do with the direct causes for the uprising in Chiapas. These signifiers were picked up and rapidly circulated through e-mail and www-sites across national and ideological boundaries' (ibid.). Indeed, most Zapatista websites are located outside Mexico, there is no evidence of a direct Zapatista presence on the web, and it is unlikely that there could be a direct presence on the web given the practical difficulties of establishing Internet connections from remote jungle locations. Instead of directly orchestrating a scale-jumping strategy, the Zapatistas have communicated their message beyond Chiapas through the connections they have built with a variety of pre-existing organizations 'in different places, with different agendas (churches, human rights groups, left political groups) that converge around the issue of the Zapatista uprising. Cyberzapatistas are everywhere, but they are not controlled by the Zapatistas in Chiapas (ibid.: 171). The Zapatistas have mobilized international support through the combination of a brilliant reframing of their message to give it resonance at broader scales, charismatic interest-provoking leadership, and a fortuitous network of connections to diverse organizations with the skills and resources to broadly disseminate their message. The Zapatistas have not, however, bridged the local/global gulf by directly speaking to the world through the Internet; the involvement of many other actors and organizations was required.

While the Zapatistas' efforts at scalar reframing are perhaps the best known, numerous other examples abound, including linking less accepted frames with more widely accepted ones, and scaling down to give broad abstract norms context-specific meaning. For instance, the international women's movement had limited success until it reframed its message as 'women's rights are human rights', thereby gaining women's rights a much higher level of international acceptance. Once women's rights were established as an international norm, the norm could be appropriated and redefined in reference to a variety of place-specific struggles. The practice of 'female circumcision', for example, has been common in many African as well as some Asian and Middle Eastern countries for centuries, but it was not until a network of human rights and women's organizations renamed the practice 'female genital mutilation' and reframed it as a violation of women's universal human rights, that opposition to that particular practice began to be effective (Keck and Sikkink, 1998: 20).

Although scalar reframing is an important means by which social movements build broad bases of support and crucial alliances, it must be recognized that spatial relations among different groups of people are not exhausted by the concept of spatial scale. Although 'social life operates in and constructs some sort of nested hierarchical space' (Smith, 1992: 73), horizontal place-to-place relations may be just as important as scalar relations in social movement mobilization. Different geometries of spatial relations are apparent in the structures of social movement organizations. Jackie Smith's (1997) extensive analysis of transnational social movement organizations (TSMOs) identifies four basic types of structure: federation (a scalar hierarchy including national, regional and/or local branches in addition to a transnational headquarters); coalition (horizontal networks of organizations); individual membership (no mediating organizational layers between the individual and the transnational organization); and professional (usually individual membership, but with the implied mediating influence of professional organizations). There are advantages and disadvantages to each type of organizational structure. The nested hierarchies of federated structures reduce costs, heighten efficiency, and provide clear lines of authority (ibid.). They make quick reactions to major events and policy decisions possible, while facilitating multi-scalar framing and issue selection, thereby improving the relationships between transnational social movements and place-specific norms, issues, and identities. The risk of such a structure is that the organization may move more quickly than discussions among movement supporters, in turn alienating the grassroots base (Chapman, 2001). Smith (1997) finds TSMOs with federated structures are active in more countries (38 on average) than other types of TSMOs. The former also have the highest number of contacts with intergovernmental organizations and non-governmental organizations (12 on average).

By comparison, TSMOs with coalition (or network) structures tend to be particularly adept at horizontal information exchange and at selecting issues and framing messages so they resonate at the scale of the organization (usually local, regional or national). However, lacking clear lines of authority, they may be slow to act and may have difficulty coordinating action when it is taken. They also tend to be active in fewer countries (28 on average) and have fewer contacts with international organizations and non-governmental organizations (8 on average) (ibid.). A variation on the network structure is the wheel structure that provides one or more focal points for information exchange (Chapman, 2001). Wheel structures are better at sorting information than webs, but may still suffer coordination and response problems. TSMOs with individual membership and professional membership structures are based on members accepting an agenda and set of values unmediated by place-specific circumstances, but in cases like Médecins Sans Frontières/Doctors without Borders regional and national professional organizations may substitute for scale-specific discussion and agenda-setting absent in the organization itself.

One of Smith's (1997) striking findings is that there has been a dramatic shift in the prevalence of TSMOs between 1973 and 1993. Federations accounted for 50% of all TSMOs in 1973; by 1993 that figure had dropped to 29%. In the same period, organizations that were part of coalitions increased from 25% to 40% of the total. Individual membership organizations increased slightly, from 20% to 22%, and professional membership organizations increased from 2% to 7%. Smith argues that the large drop in federated organizations and the parallel increase in coalitions is due to the greater availability of 'electronic mail, faxes, and other communication technologies [as well as] transnational transportation, [making] it feasible for more decentralized organizations to operate' (ibid.: 54–55). In other words, the growth of organizational networks/coalitions is in large measure a function of time–space compression. The continuing reduction in communication and transportation costs and more widespread use of email and the Internet strongly suggests this trend is continuing. Certainly, the role the Internet has played in facilitating the organization of anti-globalization protests in Seattle, Prague, Washington DC, Quebec City and elsewhere, is widely acknowledged.

There is no doubt the Internet benefits transnational social movements in a number of ways. It allows activists to store a tremendous bank of movement-collected and generated information in a readily accessible location, to be retrieved as needed. It provides activists with access to even more information from non-movement sources. It provides a means of communication that bypasses the gatekeepers of the corporate mass media. It facilitates new repertoires of protest such as direct email and Internet petition campaigns. It allows individuals with interests in a given political movement to easily locate information about that movement and quickly become involved in some way. And perhaps most importantly, it facilitates the coordination of campaigns on a global scale. The impact of the Internet on social movement mobilization has been so profound, Thu Nguyen and Alexander (1996) suggest, that we can expect place-based forms of mobilization to rapidly decline, to be replaced by cyber-mobilization.

But before welcoming the dawn of a truly transnational, disembedded, non-hierarchical, global civic realm, it is important to recognize the limitations and drawbacks of cyber-organizing. One of the underlying assumptions of many cyber-enthusiasts is that the Internet provides for nearly ubiquitous communication. But as Dodge and Kitchen (2001: 41) bluntly put it, 'cyberspace is patently not accessible to all'. Internet access exhibits substantial social and spatial differentiation:

> In fact, far from creating a more egalitarian society, many commentators have suggested that cyberspace will reproduce and reinforce the rising dual economy within countries [...] and between the developed and the developing world [...]. At present, cyberspace is only accessible to those who have the telecommunication infrastructure (a computer, a telephone line), can afford the equipment, have the skills to operate the equipment, and the time to interact with it [...]. Usage is currently dominated by people in the middle- to upper-level income bracket. (Dodge and Kitchen, 2001: 41)

This income bracket tends to be strongly correlated with the highly educated segments of the labour force that have personal and professional interests in global affairs – in other words, the cosmopolitan elite. While patterns of usage reinforce the notion of the Internet as an important technology of transnational mobilization among those with a cosmopolitan orientation, it also suggests limited mobilization efficacy among more place-bound collectivities. Further, Internet sites are self-selected by users. Indeed, the Internet provides the capacity to filter out unwanted information to such an extent that Sunstein (2002) has dubbed users' self-selected news exposure 'The Daily Me'. Sunstein is concerned with the implications of 'any situation in which thousands, or perhaps millions or even tens of millions of people, are mainly listening to louder echoes of their own voices' (cited in Evans, 2002). Of course, within a movement of already-mobilized individuals and groups, such self-selection of information actually produces a greater depth of knowledge and efficiency of communication. But if, as Michael Albert argues in the statement that opens this chapter, the objective is to build a broad base of support for the anti-globalization movement (and other emancipatory movements), movement messages must reach beyond the core of committed activists. 'Weak' supporters and uncommitted 'bystander publics' must be mobilized (Ennis and Schreuer, 1987). To do this, movement messages must reach their target audiences and be framed in ways that will resonate with specific collectivities 'where they live and in accord with their dispositions and possibilities' (Albert, 2001a: 2).

Within the anti-globalization movement in particular there is an increasingly widespread recognition that greater emphasis needs to be placed on a balanced global/local mobilization dialectic. To date, the movement has primarily focused on mobilizing core activists who will travel 'to distant cities and [demonstrate] in the midst of clubs and tear gas' (ibid.). This strategy, which has demonstrated the ability of the movement to negotiate space on a par with global capital and global governance institutions, is gradually giving way to a strategy that empha-sizes building stronger, broader bases of support in place. In a position paper circulated before the June 2002 Kananaskis G8 summit (held just outside of Calgary, Alberta), Yutaka Dirks called for 'Local Resistance: Not Summit Hopping!':

We must begin looking to our communities and neighborhoods to build resis-tance 'at home.' Instead of hopping on a plane to take part in actions outside of a meeting of the ruling elite, we should be identifying who benefits from their decisions and policies, who pays the price so that these people can benefit, and how we can make principled connections with those people and movements who are already fighting against their oppression in our own communities. (Dirks, 2002: 1)

Similarly, Spector (2002) argues that 'we have to immerse ourselves in the lives of the people we hope to "win over" to a clearer understanding of the nature of the capitalist-imperialist system [...] it is in our personal relationships that we can build [...] trust and understanding.'

Conclusion

The anti-globalization movement, and transnational social movements gener-ally, are beginning to develop mobilization strategies that consider both the need for global communication and the need for resonance among place-specific collectivities. These are not the same. While the Internet provides an effective means of global communication and coordination, it is far from the panacea its boosters suggest. Its fragmented and self-selected cosmopolitan audiences, its non-hierarchical structure that inhibits (but not necessarily precludes) scale-sensitive framing, and the generally weak bonds that Internet communications produce (Kitchen, 1998), all suggest severe limitations to cyberspace mobilization. Indeed, social movement organizations coordinated through the horizontal links of the Internet appear to be less effective than federated organizations at building transnational support. On the other hand, the Internet has proven to be a very valuable means of communicating and coordinating action across space, bypassing the gatekeepers of the corporate mass media.

While attention to global actors and institutions is imperative in an increasingly globalized world, going global is no more a panacea for social movements than is the localism that characterizes many forms of resistance. Globalization has resulted in greater international integration and a growing cosmopolitanism, but at the same time most people continue to live their lives in place, and their collective identities and values are shaped there too. Imagined communities such as the nation-state, moreover, remain a powerful dimension of collective identity, and not to be overlooked are the many isolated individuals lacking strong collective identification. Diverse and fluctuating forms of collective identity call into question broad master frames for transnational mobilization, whether based in class or culture. Instead, transnational social movements that hope to mobilize support beyond a core of committed activists must reach potential supporters by appealing to the experiences, identities and meanings that have shaped their lives. Broad transnational mobilization demands continuous attention to the dialectical relationship between abstract principles of material and cultural justice and the specific circumstances, identities, and relationships of lives lived in real places.

References

Agnew, J. (1996) 'Mapping Politics: How Context Counts in Electoral Geography', *Political Geography*, 15 (2): 129–46.

Albert, M. (2001a) 'New Targets', <http://www.zmag.org/CrisesCurEvts/Globalism/new_targets.htm>

Albert, M. (2001b) 'The Movements against Neoliberal Globalisation from Seattle to Porto Allegre', <http://www.zmag.org/albertgreecctalk.htm>.

Anderson, B. (1991) *Imagined Communities*. London: Verso.

Appadurai, A. (1993) 'Disjuncture and Difference in the Global Cultural Economy', in B. Robbins (ed.), *The Phantom Public Sphere*. Minneapolis: University of Minnesota Press, pp. 269–96.

Barber, B. (1995) *Jihad vs. McWorld*. New York: Time Books.

Boli, J. and J. Thomas (eds) (1999) *Constructing World Culture*. Stanford, CA: Stanford University Press.

Calhoun, C. (1988) 'The Radicalism of Tradition and the Question of Class Struggle', in M. Taylor (ed.), *Rationality and Revolution*. New York: Cambridge University Press, pp. 129–75.

Castells, M. (1996) *The Rise of the Network Society*. Oxford: Blackwell.

Chapman, J. (2001) 'What Makes International Campaigns Effective? Lessons from India and Ghana', in M. Edwards and J. Gaventa (eds), *Global Citizen Action*. Boulder, CO: Lynne Rienner, pp. 259–74.

Cheah, P. and B. Robbins (eds) (1998) *Cosmopolitics*. Minneapolis: University of Minnesota Press.

Cox, K. and A. Mair (1988) 'Locality and Community in the Politics of Local Economic Development', *Annals of the Association of American Geographers*, 78 (2): 307–25.

Delaney, D. and H. Leitner (1997) 'The Political Construction of Scale', *Political Geography*, 16 (2): 93–7.

Dicken, P., J. Peck and A. Tickell (1997) 'Unpacking the Global', in R. Lee and J. Willis (eds), *Geographies of Economies*. London and New York: Arnold, pp. 158–66.

Dirks, Y. (2002) 'Doing Things Differently this Time: Kananaskis G8 Meeting and Movement Building', <http://g8.activist.ca/calltoaction/local.html>.

Dodge, M. and R. Kitchen (eds) (2001) *Mapping Cyberspace*. London and New York: Routledge.

Edwards, M. and J. Gaventa (eds) (2001) *Global Citizen Action*. Boulder, CO: Lynne Rienner.

Ennis, J. and R. Schreuer (1987) 'Mobilizing Weak Support for Social Movements: The Role of Grievance, Efficacy, and Cost', *Social Forces*, 66 (2): 390–409.

Escobar, A. (2001) 'Culture Sits in Places: Reflections on Globalism and Subaltern Strategies of Localization', *Political Geography*, 20: 139–74.

Evans, D. (2002) 'Plugged in to e-Democracy', *The Calgary Herald*, 14 October: D5.

Featherstone, M. (ed.) (1990) *Global Culture*. London: Sage.

Fraser, N. (1995) 'From Redistribution to Recognition? Dilemmas of Justice in a "Post-Socialist" Age', *New Left Review*, 212: 68–93.

Froehling, O. (1999) 'Internauts and Guerrilleros: The Zapatista Rebellion in Chiapas, Mexico and its Extension into Cyberspace', in M. Crang, P. Crang and J. May (eds), *Virtual Geographies*. London: Routledge, pp. 164–77.

Gitlin, T. (1995) *The Twilight of Common Dreams: Why America is Wracked by Culture Wars*. New York: Metropolitan Books.

Habermas, J. (1984) *The Theory of Communicative Action* (Volume 1). Boston, MA: Beacon Press.

Habermas, J. (1987) *The Theory of Communicative Action* (Volume 2). Boston, MA: Beacon Press.

Hall, S. (1997) 'Old and New Identities, Old and New Ethnicities', in A. King (ed.), *Culture, Globalisation and the World-System*. Minneapolis: University of Minnesota Press, pp. 41–68.

Hannerz, U. (1992) *Cultural Complexity*. New York: Columbia University Press.

Hannerz, U. (1996) *Transnational Connections: Culture, People, Places*. New York and London: Routledge.

Harvey, D. (1989) *The Condition of Postmodernity*. Oxford and Cambridge, MA: Blackwell.

Harvey, D. (1996) *Justice, Nature, and the Geography of Difference*. Oxford: Blackwell.

Harvey, D. (2000) *Spaces of Hope*. Berkeley and Los Angeles: University of California Press.

Hobsbawm, E. (1996) 'Identity Politics and the Left', *New Left Review*, 217: 38–47.

Honneth, A. and J. Anderson (1996) *The Struggle for Recognition*. Cambridge, MA: MIT Press.

Hudson, R. (2001) *Producing Places*. New York: Guilford Press.

Jencks, C. (1979) 'The Social Basis of Unselfishness', in H. Gans, N. Glazer, J. Gusfield and C. Jencks (eds), *On the Making of Americans*. Philadelphia: University of Pennsylvania Press, pp. 63–86.

Jessop, B. (1994) 'Post-Fordism and the State', in A. Amin (ed.), *Post-Fordism: A Reader*. Oxford: Blackwell, pp. 251–79.

Keck, M. and K. Sikkink (eds) (1998) *Activists Beyond Borders*. Ithaca, NY and London: Cornell University Press.

Khagram, S., J. Riker and K. Sikkink (2002) 'From Santiago to Seattle: Transnational Advocacy Groups Restructuring World Politics', in S. Khagram, J. Riker and K. Sikkink (eds), *Transnational Social Movements, Networks, and Norms*. Minneapolis: University of Minnesota Press, pp. 3–23.

King, A. (ed.) (1997) *Culture, Globalisation and the World-System*. Minneapolis: University of Minnesota Press.

Kitchen, R. (1998) 'Towards Geographies of Cyberspace', *Progress in Human Geography*, 22: 385–406.

Kobayashi, A. (1997) 'The Paradox of Difference and Diversity (or, Why the Threshold Keeps Moving)', in J. P. Jones, H. Nast and S. Roberts (eds), *Thresholds in Feminist Geography*. Lanham, MD: Rowman and Littlefield, pp. 3–9.

Mansbridge, J. (1990) 'On the Relation of Altruism and Self-Interest', in J. Mansbridge (ed.), *Beyond Self-Interest*. Chicago: University of Chicago Press, pp. 133–46.

Marden, P. (1997) 'Geographies of Dissent: Globalisation, Identity and the Nation', *Political Geography*, 16: 37–64.

Markusen, A. (1987) *Regions: The Economics and Politics of Territory*. Totowa, NJ: Rowman and Littlefield.

Massey, D. (1994) 'A Global Sense of Place', in *Space, Place, and Gender*. Minneapolis: University of Minnesota Press, pp. 146–56.

McAdam, D. (1999) *Political Process and the Development of Black Insurgency, 1930–1970*. Chicago: University of Chicago Press.

McAdam, D., S. Tarrow and C. Tilly (2001) *Dynamics of Contention*. Cambridge: Cambridge University Press.

Melucci, A. (1994) 'A Strange Kind of Newness: What's 'New' in New Social Movements', in E. Larana, H. Johnston and J. Gusfield (eds), *New Social Movements: From Ideology to Identity*. Philadelphia, PA: Temple University Press, pp. 101–30.

Meyer, B. and P. Geschiere (1999) 'Globalisation and Identity: Dialectics of Flow and Closure', in B. Meyer and P. Geschiere (eds), *Globalisation and Identity: Dialectics of Flow and Closure*. Oxford: Blackwell, pp. 1–15.

Meyer, J., J. Boli, G. Thomas and F. Ramirez (1997) 'World-Society and the Nation-State', *American Journal of Sociology*, 103: 144–81.

Miller, B. (1992) 'Collective Action and Rational Choice: Place, Community, and the Limits to Individual Self-Interest', *Economic Geography*, 68: 22–42.

Miller, B. (2000) *Geography and Social Movements*. Minneapolis: University of Minnesota Press.

Mittelman, J. (2000) *The Globalisation Syndrome: Transformation and Resistance*. Princeton, NJ: Princeton University Press.

Olson, M. (1965) *The Logic of Collective Action*. Cambridge, MA: Harvard University Press.

Routledge, P. (1998) 'Going Globile: Spatiality, Embodiment, and Mediation in the Zapatista Insurgency', in G. O'Tuathail and S. Dalby (eds), *Rethinking Geopolitics*. London and New York: Routledge, pp. 240–60.

Smith, D. (2000) 'Social Justice Revisited', *Environment and Planning A*, 32: 1149–62.

Smith, J. (1997) 'Characteristics of the Modern Transnational Social Movement Sector', in J. Smith, C. Chatfield and R. Pagnucco (eds), *Transnational Social Movements and Global Politics*. Syracuse, NY: Syracuse University Press, pp. 42–58.

Smith, J., C. Chatfield and R. Pagnucco (eds) (1997) *Transnational Social Movements and Global Politics*. Syracuse, NY: Syracuse University Press.

Smith, N. (1992) 'Geography, Difference, and the Politics of Scale', in J. Doherty, E. Graham and M. Malek (eds), *Postmodernism and the Social Sciences*. London: Macmillan, pp. 57–79.

Smith, N. (1993) 'Homeless/Global: Scaling Places', in J. Bird, B. Curtis, T. Putnam, G. Robertson and L. Tickner (eds), *Mapping the Futures: Local Cultures, Global Change*. New York: Routledge, pp. 87–119.

Snow, D. and R. Benford (1988) 'Ideology, Frame Resonance and Participant Mobilization', in B. Klandermans, H. Kriesi and S. Tarrow (eds), *International Social Movement Research*. Vol. 1: *From Structure to Action: Comparing Social Movements Research Across Cultures*. Greenwich, CT: JAI Press, pp. 197–217.

Spector, A. (2002) 'Whose Media? Whose Movement?', Posting to SOCIAL-MOVE-MENTS@LISTSERV.HEANET.IE, <http://listserv.heanet.ie/social-movements.html>, 8 October.

Sunstein, C. (2002) *Republic.com*. Princeton, NJ: Princeton University Press.

Swyngedouw, E. (1997a) 'Excluding the Other: The Production of Scale and Scaled Politics', in R. Lee and J. Willis (eds), *Geographies of Economies*. London and New York: Arnold, pp. 167–76.

Swyngedouw, E. (1997b) 'Neither Global nor Local: "Glocalization" and the Politics of Scale', in K. Cox (ed.), *Spaces of Globalisation*. New York and London: Guilford Press, pp. 137–66.

Tarrow, S. (1998) *Power in Movement*. Cambridge: Cambridge University Press.

Thu Nguyen, D. and J. Alexander (1996) 'The Coming of Cyberspace Time and the End of Polity', in R. Shields (ed.), *Cultures of Internet: Virtual Spaces, Real Histories and Living Bodies*. London: Sage, pp. 99–124.

Tilly, C. (1978) *From Mobilization to Revolution*. New York: Random House.

Tilly, C. (2003) 'Contention over Space and Place', *Mobilization*, 8(2): 221–5.

Weiss, L. (1998) *The Myth of the Powerless State*. Ithaca, NY: Cornell University Press.

Index

QM LIBRARY (MILE END)

WITHDRAWN
FROM STOCK
QMUL LIBRARY